Introduction:

This book is intended for students studying Introductory Structural Analysis.

The Author has strived to present problems that would be found in a typical engineering class, in a hand drawn style that will be familiar to any student who has put pencil to engineering paper.

If any errors or omissions are discovered, please report them to NieKo Technical Publishing for correction in future editions.

Book Notation:

A - Area (sq. in.)

L – Length (feet or inches)

E – Modulus of Elasticity (K/in2) or (ksi)

I – Moment of Inertia (in4)

k – kips = 1000 lbs.

mv – Virtual Moment (kft)

M – Real Moment (kft)

n – Virtual Axial Force (k)

N – Real Axial Force (k)

FEM – Fixed End Moment (kft)

DF – Distribution Factors

FBD – Free Body Diagram

f aa – Flexibility Coefficient

Δ - Deflection (in)

Table of Contents

<u>Book Standards</u>

$\underline{\underline{300^k}}$ — Double line denotes a final answer.
— Single line denotes an intermediate answer.

$\underline{\underline{300^k}}$ $\rightarrow$ direction sign shows positive direction of force or moment

<u>Possible</u>

3 reactions ⊥⁄⁄⁄ — denotes a fixed support

2 reactions △ — denotes a pin support ⎫ Idealized Support
                                        ⎬ Conditions
1 reaction ⊙⁄⁄⁄ — denotes a roller support ⎭

Equations of equilibrium 3 in X-y Plane (2d)

$\uparrow \Sigma F_y$ — Sum of y forces — positive upwards

$\rightarrow \Sigma F_x$ — Sum of x forces — positive to the right

$\circlearrowleft \Sigma M_x$ — Sum of moments — positive direction is Counter Clockwise about location "x"

A   B

- beams, frame members, or truss members are depicted with a single line.
- beam shown has a pin at A and a roller at B.
- length of beam is from A to B

(Comments) — hopefully helpful comments

Since the problems shown deal with 2 dimensional structures, the following equations of equilibrium are used throughout this book.

Sum of the forces in the X direction are zero.

$$\Sigma Fx = 0$$

Sum of the forces in the Y direction are zero.

$$\Sigma Fy = 0$$

Sum of the moments about the Z axis are zero.

$$\Sigma M = 0$$

Beams

Beam 1

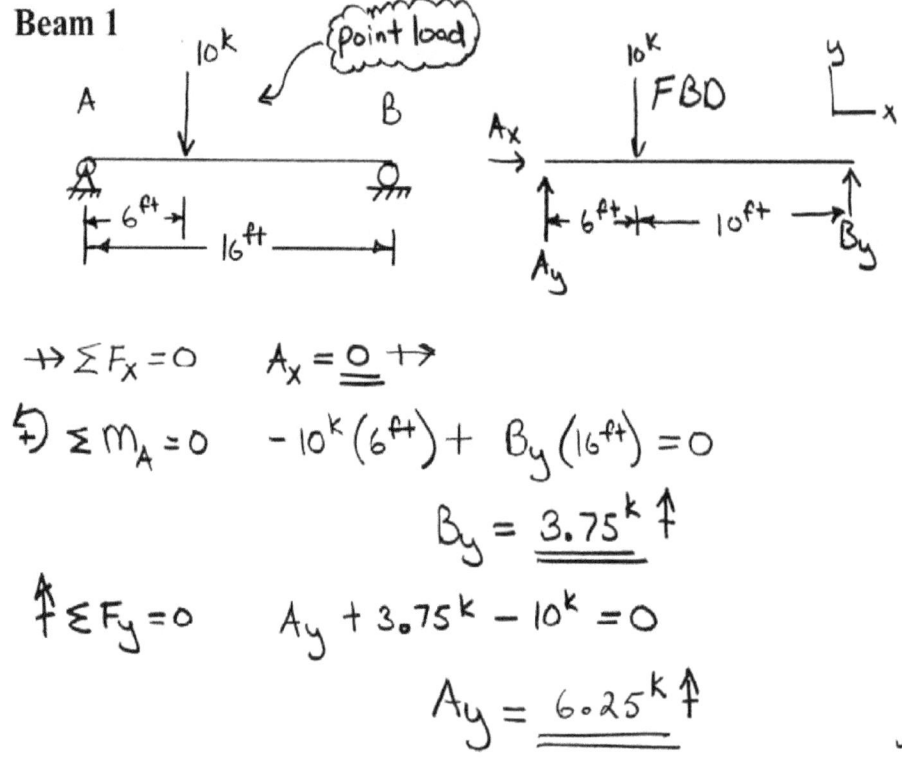

$$\twoheadrightarrow \Sigma F_x = 0 \qquad A_x = \underline{\underline{0}} \;\twoheadrightarrow$$

$$\circlearrowleft \Sigma M_A = 0 \qquad -10^k(6^{ft}) + B_y(16^{ft}) = 0$$

$$B_y = \underline{\underline{3.75^k}} \uparrow$$

$$\uparrow \Sigma F_y = 0 \qquad A_y + 3.75^k - 10^k = 0$$

$$A_y = \underline{\underline{6.25^k}} \uparrow$$

Beam 2

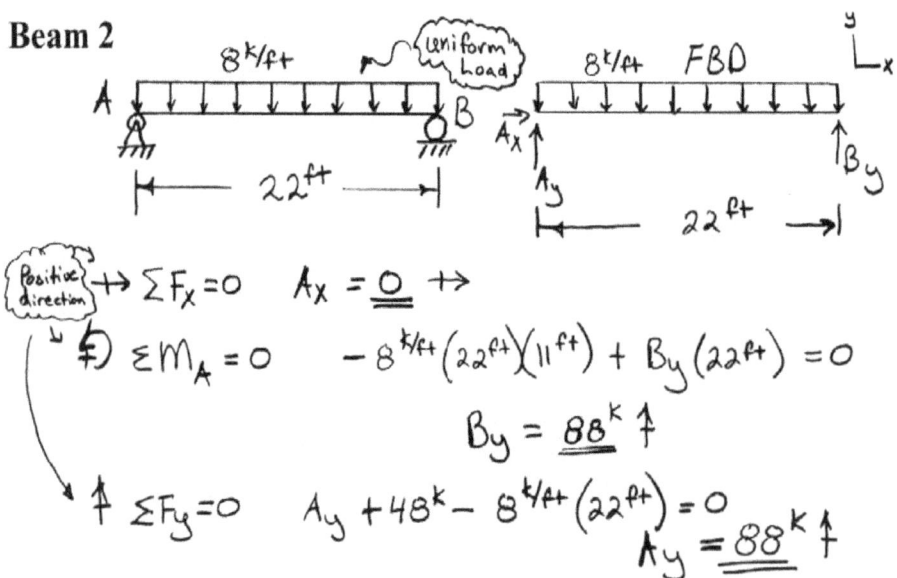

$$\twoheadrightarrow \Sigma F_x = 0 \qquad A_x = \underline{\underline{0}} \;\twoheadrightarrow$$

Positive direction

$$\circlearrowleft \Sigma M_A = 0 \qquad -8^{k/ft}(22^{ft})(11^{ft}) + B_y(22^{ft}) = 0$$

$$B_y = \underline{\underline{88^k}} \uparrow$$

$$\uparrow \Sigma F_y = 0 \qquad A_y + 48^k - 8^{k/ft}(22^{ft}) = 0$$

$$A_y = \underline{\underline{88^k}} \uparrow$$

Beam 3

A $12^{k}/_{ft}$ 15^{k}

← uniform load plus point load

Cantilever Beam

$\longleftarrow 15^{ft} \longrightarrow$

A fixed support can handle X and y forces plus a moment

$\underset{x}{\overset{y}{\llcorner}}$

FBD

$12^{k}/_{ft}$ 15^{k}

A_x

M_A A_y

$\longleftarrow 15^{ft} \longrightarrow$

Moment at A

$\xrightarrow{+} \Sigma F_x = 0 \quad A_x = \underline{0} \rightarrow$

$\uparrow^{+} \Sigma F_y = 0 \quad -12^{k/ft}(15^{ft}) - 15^{k} + A_y = 0$

$$A_y = \underline{\underline{195^{k}}} \uparrow$$

$\overset{+}{\curvearrowleft} \Sigma M_A = 0 \quad -12^{k/ft}(15^{ft})(7.5^{ft}) - 15^{k}(15^{ft}) + M_A = 0$

$$M_A = \underline{\underline{1575^{kft}}} \overset{+}{\curvearrowright}$$

Note: To find the moment of a uniform load, make a resultant force at the centroid of the load. (In pictures)

$12^{k}/ft$

M_A

$\Rightarrow$

M_A $12^{k}/ft(15^{ft}) = 180^{k}$

$\longleftarrow 7.5^{ft} \longrightarrow$

$M_A = 12^{k/ft}(15^{ft})(7.5^{ft})$ or $M_A = 180^{k}(7.5^{ft})$

Beam 4

$$\rightarrow \Sigma F_X = 0 \qquad A_X = \underline{\underline{0}}$$

Summing moments around A is easier than B.

$$\circlearrowleft \Sigma M_A = 0$$

$$-5^k(8^{ft}) - 12^k(24^{ft}) - 9^{k/ft}(24^{ft})(12^{ft}) + B_y(16^{ft}) = 0$$

$$B_y = \underline{\underline{182.5^k}}$$

$$\uparrow \Sigma F_y = 0 \qquad A_y - 5^k - 12^k - 9^{k/ft}(24^{ft}) + 182.5^k = 0$$

$$A_y = \underline{\underline{50.5^k}}$$

If you had summed moments around B instead of A, your equation would be:

$$\circlearrowleft \Sigma M_B = 0$$

$$0 = -A_y(16^{ft}) + 5^k(8^{ft}) + 9^{k/ft}(16^{ft})(8^{ft}) - 12^k(8^{ft}) - 9^{k/ft}(8^{ft})(4^{ft})$$

$$A_y = \underline{\underline{50.5^k}}$$

Beam 5

10 $^{kN}/m$

C

A

B

6m   4,5m

Find the reactions
A - roller
B - Pin
C - fixed

FBD

10 $^{kN}/m$

A

$\to B_x$

$\uparrow A_y$

B_y

6m

10 kN/m

m_c

$\gets$

$\to C_x$

B_y $\uparrow C_y$

4.5m

Left side

$\to \Sigma F_x = 0 \quad B_x = \underline{\underline{0}}$

$\circlearrowleft \Sigma M_A = 0$

$B_y(6m) - \frac{1}{2}(10^{kN/m})(6m)(\frac{2}{3})(6m) = 0$

$B_y = \underline{\underline{20^{kN}}} \uparrow$

$\uparrow \Sigma F_y = 0$

$A_y + 20^{kN} - \frac{1}{2}(10^{kN/m})(6m) = 0$

$A_y = \underline{\underline{10^{kN}}} \uparrow$

Right Side

$\to \Sigma F_x = 0$
$C_x = \underline{\underline{0}} \quad (B_x = 0)$

$\uparrow \Sigma F_y = 0 \quad B_y = -20^k \begin{pmatrix} \text{from} \\ \text{Left} \\ \text{Side} \end{pmatrix}$

$-20^{kN} - \frac{1}{2}(10^{kN/m})(4.5m) + C_y = 0$

$C_y = \underline{\underline{42.5^{kN}}} \uparrow$

$\circlearrowleft \Sigma M_C = 0$

$\frac{1}{2}(10^{kN/m})(4.5m)(\frac{2}{3})(4.5m) + 20^{kN}(4.5m) - M_C = 0$

$M_C = \underline{\underline{67.5^{kNm}}} \circlearrowleft$

Find the reactions at the supports.

Frame 1

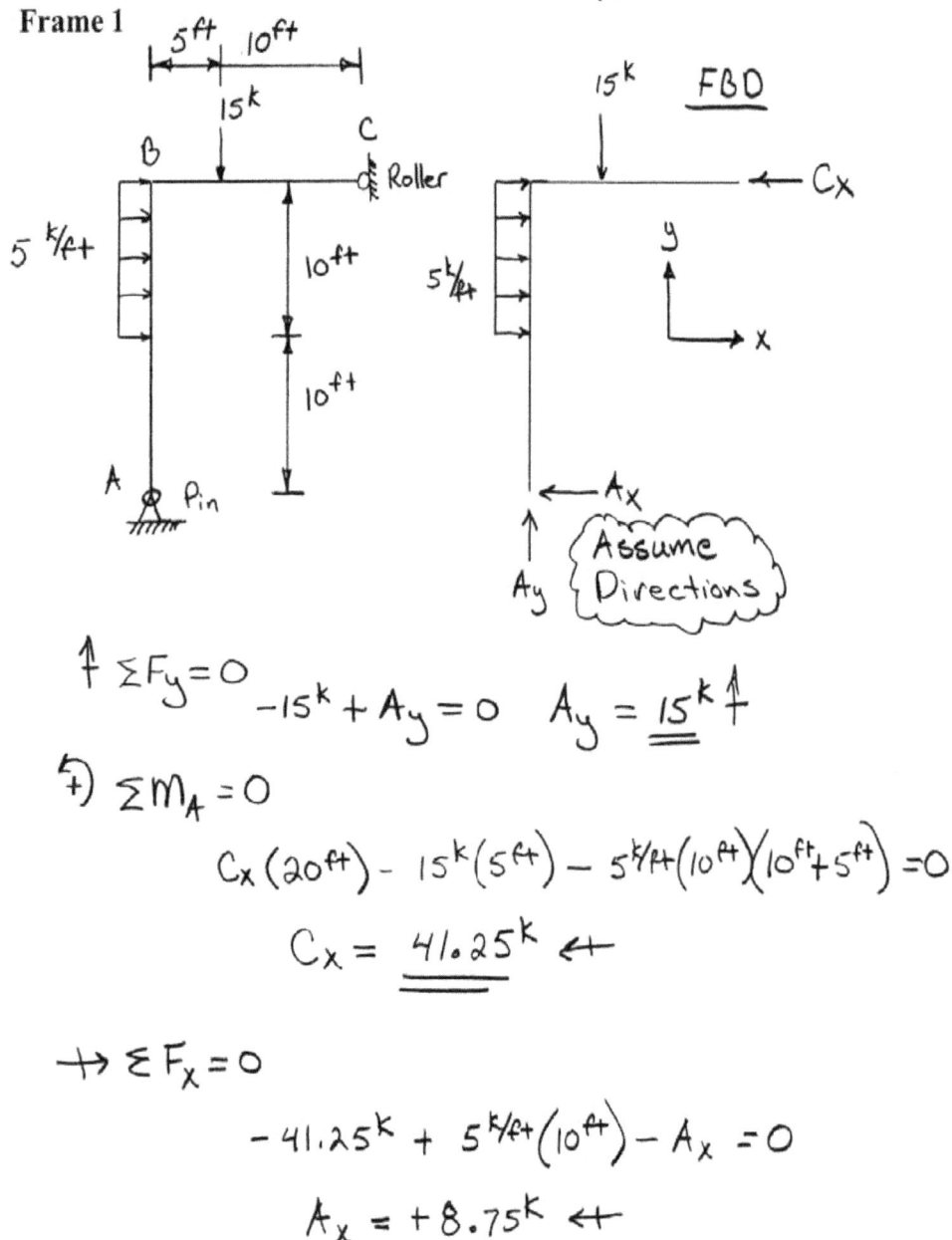

$$\uparrow \Sigma F_y = 0 \quad -15^k + A_y = 0 \quad A_y = \underline{\underline{15^k}} \uparrow$$

$$\overset{+}{\curvearrowright} \Sigma M_A = 0$$

$$C_x (20^{ft}) - 15^k (5^{ft}) - 5^{k/ft}(10^{ft})(10^{ft} + 5^{ft}) = 0$$

$$C_x = \underline{\underline{41.25^k}} \leftarrow$$

$$\twoheadrightarrow \Sigma F_x = 0$$

$$-41.25^k + 5^{k/ft}(10^{ft}) - A_x = 0$$

$$A_x = \underline{\underline{+8.75^k}} \leftarrow$$

Truss 1

Find the axial forces in members AB, BC, BI.

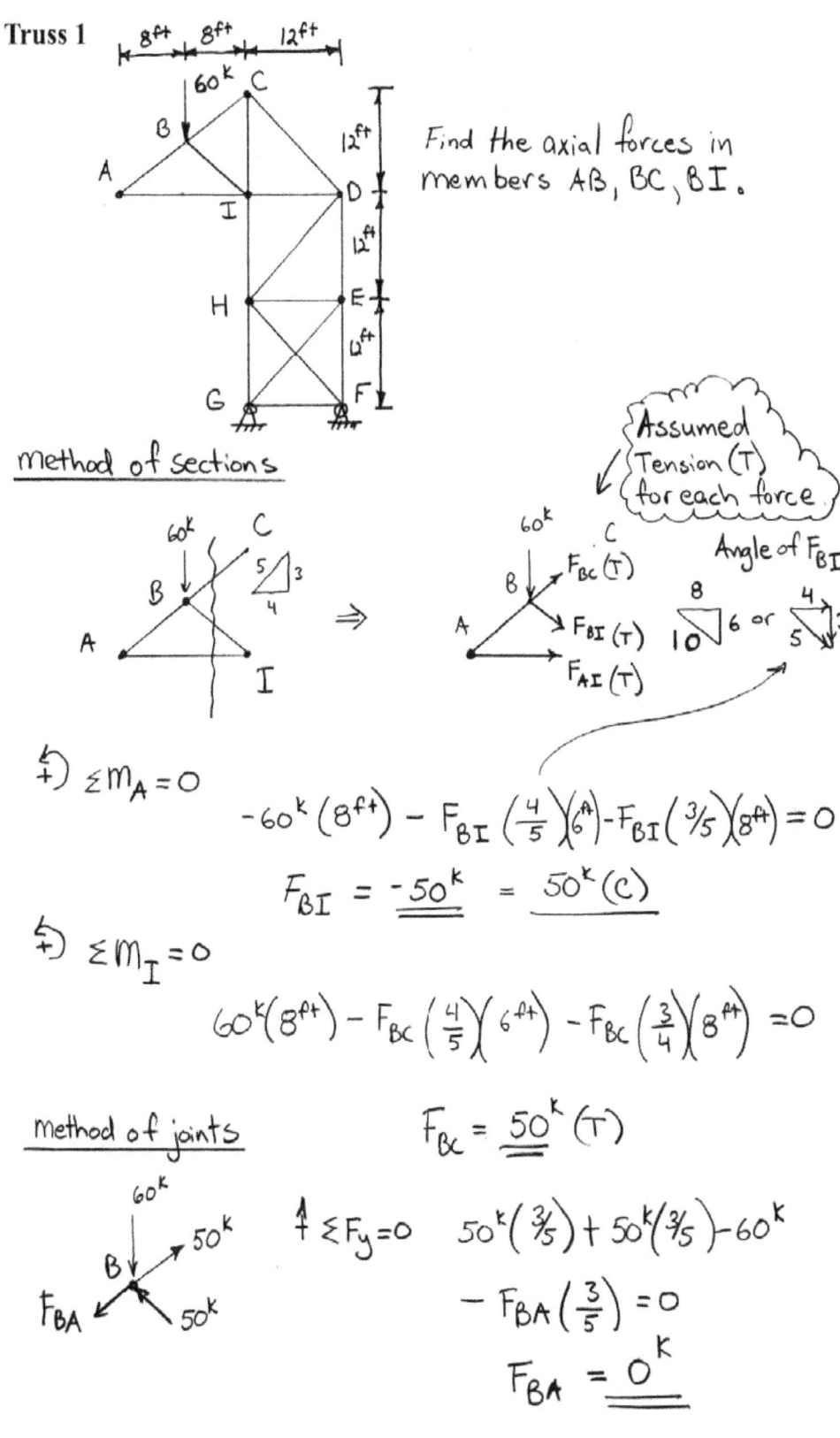

method of sections

Assumed Tension (T) for each force

Angle of F_{BI}

$\curvearrowright$ $\Sigma M_A = 0$

$$-60^k(8^{ft}) - F_{BI}\left(\frac{4}{5}\right)(6^A) - F_{BI}\left(\frac{3}{5}\right)(8^A) = 0$$

$$F_{BI} = -50^k = 50^k(C)$$

$\curvearrowright$ $\Sigma M_I = 0$

$$60^k(8^{ft}) - F_{BC}\left(\frac{4}{5}\right)(6^{ft}) - F_{BC}\left(\frac{3}{4}\right)(8^{ft}) = 0$$

method of joints

$$F_{BC} = 50^k (T)$$

$\uparrow \Sigma F_y = 0$ $\quad 50^k\left(\frac{3}{5}\right) + 50^k\left(\frac{3}{5}\right) - 60^k$

$$- F_{BA}\left(\frac{3}{5}\right) = 0$$

$$F_{BA} = 0^k$$

Method of Joints

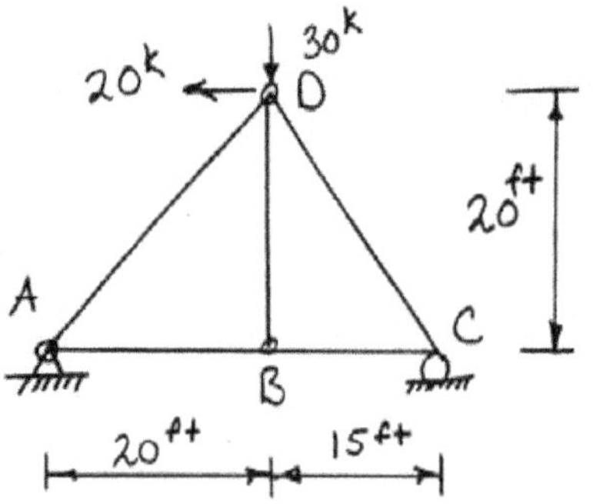

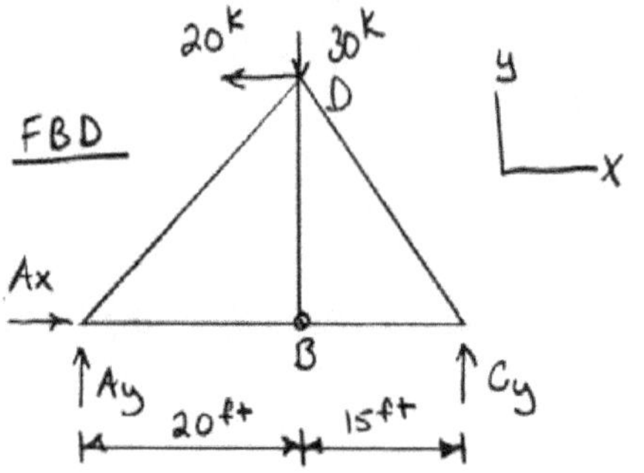

Reactions $\rightarrow \Sigma F_x = 0$

$$-20^k + A_y = 0 \quad A_y = 20^k \rightarrow$$

$\stackrel{+}{)} \Sigma M_A = 0$

$$C_y(35ft) + 20^k(20ft) - 30^k(20ft) = 0$$

$$C_y = 5.71^k \uparrow$$

$\uparrow \Sigma F_y = 0$

$$A_y - 30^k + 5.71^k = 0$$

$$A_y = 24.28^k \uparrow$$

Joint A — Assume tension

Angles: $\sqrt{800}$, $20ft$, $20ft$

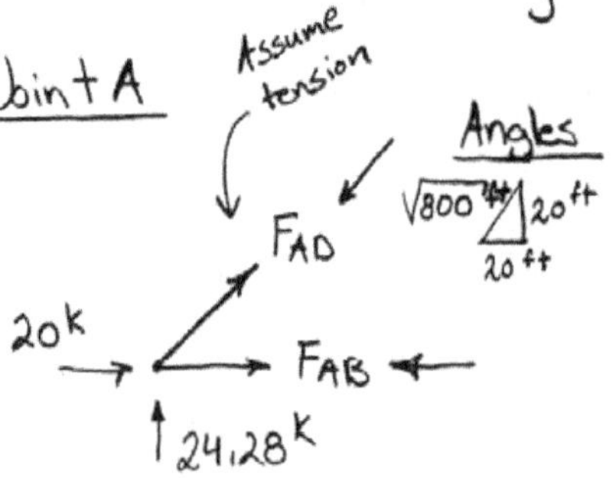

$\uparrow \Sigma F_y = 0$

$$24.28 + F_{AD}\left(\frac{20}{\sqrt{800}}\right) = 0$$

$$F_{AD} = -34.34^k (Comp.)$$

$\rightarrow \Sigma F_x = 0$

$$20^k - 34.34^k\left(\frac{20}{\sqrt{800}}\right) + F_{AB} = 0$$

$$F_{AB} = 4.28^k (Tension)$$

Joint B

$$\uparrow \Sigma F_y = 0$$
$$F_{BD} = \underline{\underline{0}}$$

$$\twoheadrightarrow \Sigma F_x = 0$$
$$F_{BC} - 4.28^k = 0$$
$$F_{BC} = \underline{\underline{4.28^k}} \, (T)$$

Joint C

Angles

$$\uparrow \Sigma F_y = 0$$

$$5.71^k + F_{CD}\left(\frac{20}{\sqrt{625'}}\right) = 0$$

$$F_{CD} = \underline{\underline{-7.13^k}} \, (c)$$

Joint D — work check

$$\uparrow \Sigma F_y = 0$$

$$-30^k + 7.13^k\left(\frac{20}{\sqrt{625'}}\right) + 34.34^k\left(\frac{20}{\sqrt{800'}}\right) = 0$$

$$\underline{\underline{-.06 = 0}} \; \checkmark \begin{pmatrix} \text{rounding} \\ \text{error} \end{pmatrix}$$

$$\twoheadrightarrow \Sigma F_x = 0$$
$$-20^k + 34.34^k\left(\frac{20}{\sqrt{800'}}\right) - 7.13\left(\frac{15}{\sqrt{625'}}\right) = 0$$

$$\underline{\underline{.004 = 0}} \; \checkmark$$

Truss 3

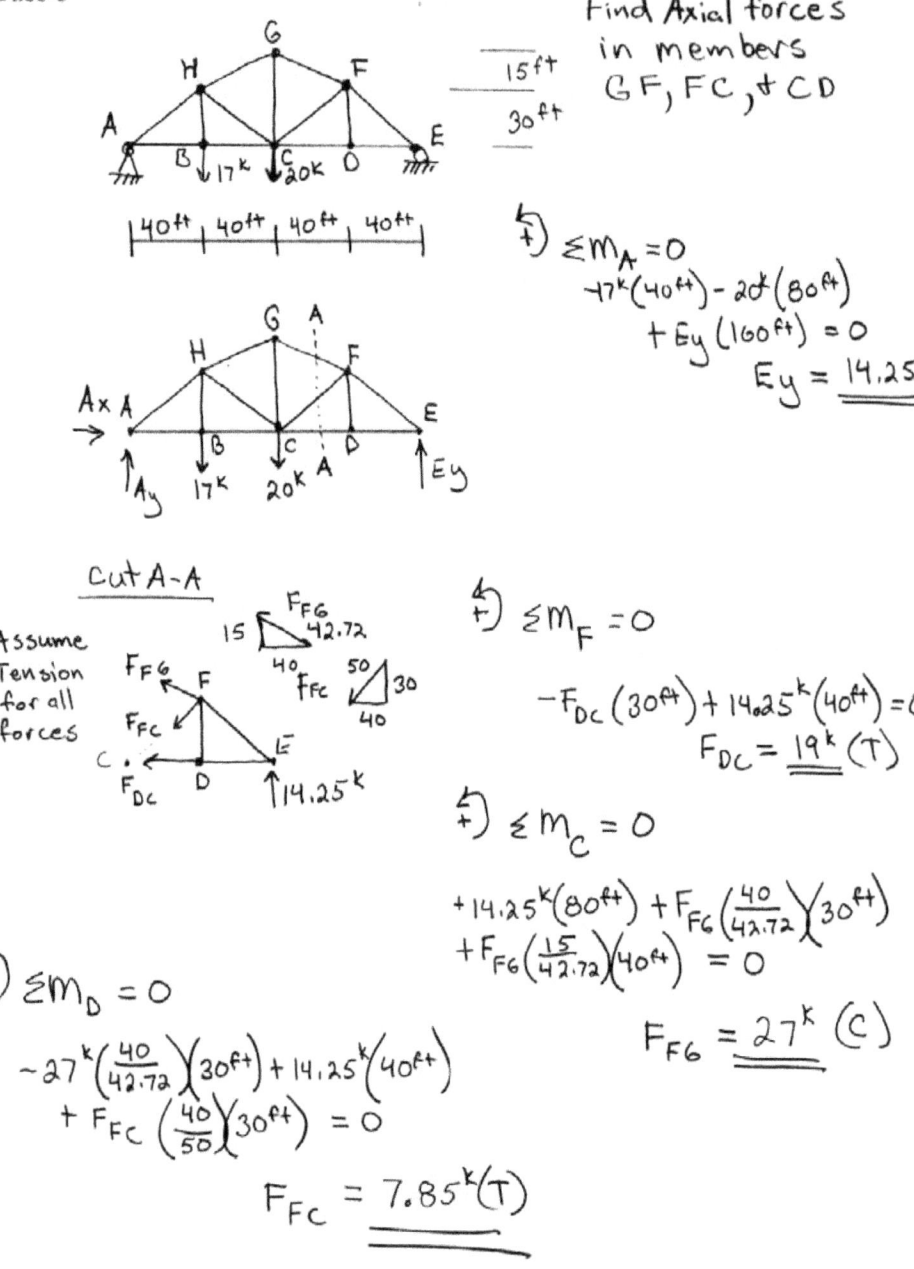

Find Axial forces
in members
$GF, FC, \& CD$

$\overline{}$ 15 ft
30 ft

$\overset{\curvearrowleft}{+})\ \Sigma M_A = 0$

$-17^k(40^{ft}) - 20^t(80^{ft})$
$\quad + E_y(160^{ft}) = 0$
$\qquad E_y = \underline{\underline{14.25^k}}$

cut A-A

Assume
Tension
for all
forces

$\overset{\curvearrowleft}{+})\ \Sigma M_F = 0$

$-F_{DC}(30^{ft}) + 14.25^k(40^{ft}) = 0$
$\qquad F_{DC} = \underline{\underline{19^k}}\ (T)$

$\overset{\curvearrowleft}{+})\ \Sigma M_C = 0$

$+14.25^k(80^{ft}) + F_{FG}\left(\frac{40}{42.72}\right)(30^{ft})$
$+F_{FG}\left(\frac{15}{42.72}\right)(40^{ft}) = 0$

$\qquad F_{FG} = \underline{\underline{27^k}}\ (C)$

$\overset{\curvearrowleft}{+})\ \Sigma M_D = 0$

$-27^k\left(\frac{40}{42.72}\right)(30^{ft}) + 14.25^k(40^{ft})$
$\quad + F_{FC}\left(\frac{40}{50}\right)(30^{ft}) = 0$

$\qquad\qquad F_{FC} = \underline{\underline{7.85^k(T)}}$

<u>Hints</u> –   Shear + Moment.

The shear diagram slope is equal to the negative
distributed load value.

ie)

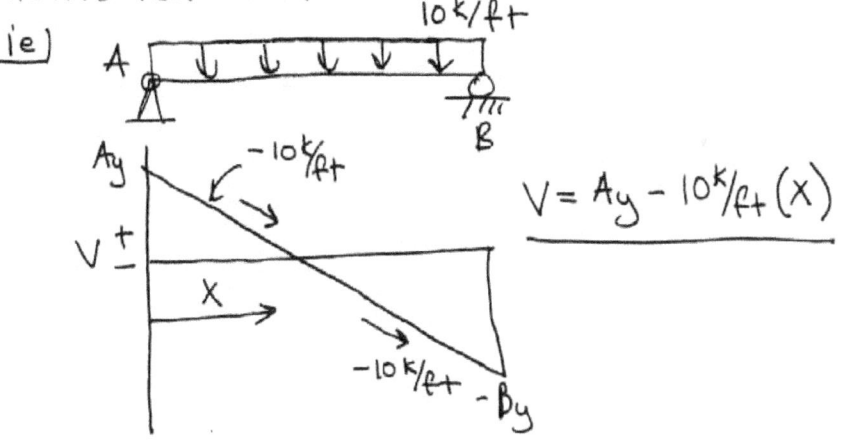

$$V = A_y - 10^k/ft\,(x)$$

The moment diagram slope is equal
to the value of the shear diagram
at that point.

ie)

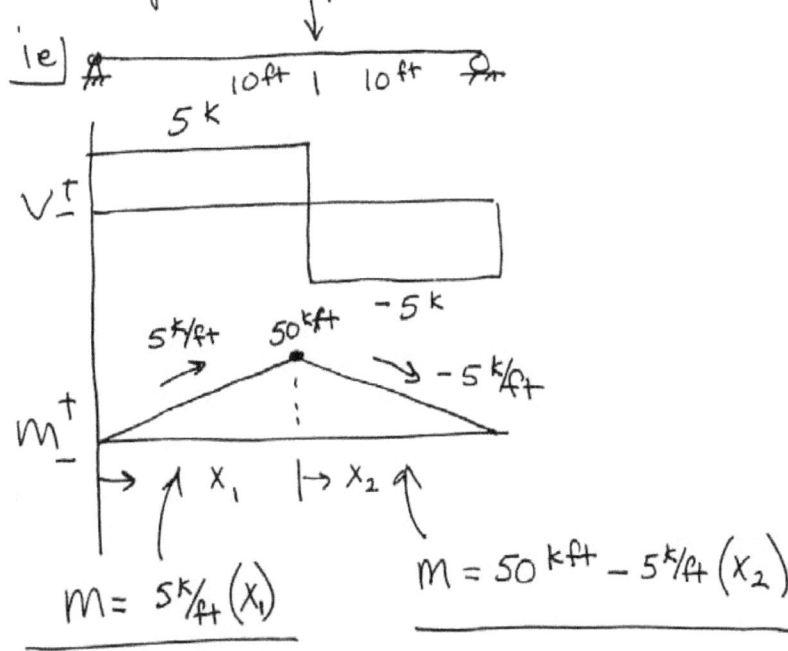

$$M = 5^k/ft\,(x_1) \qquad\qquad M = 50\,kft - 5^k/ft\,(x_2)$$

Beam 1

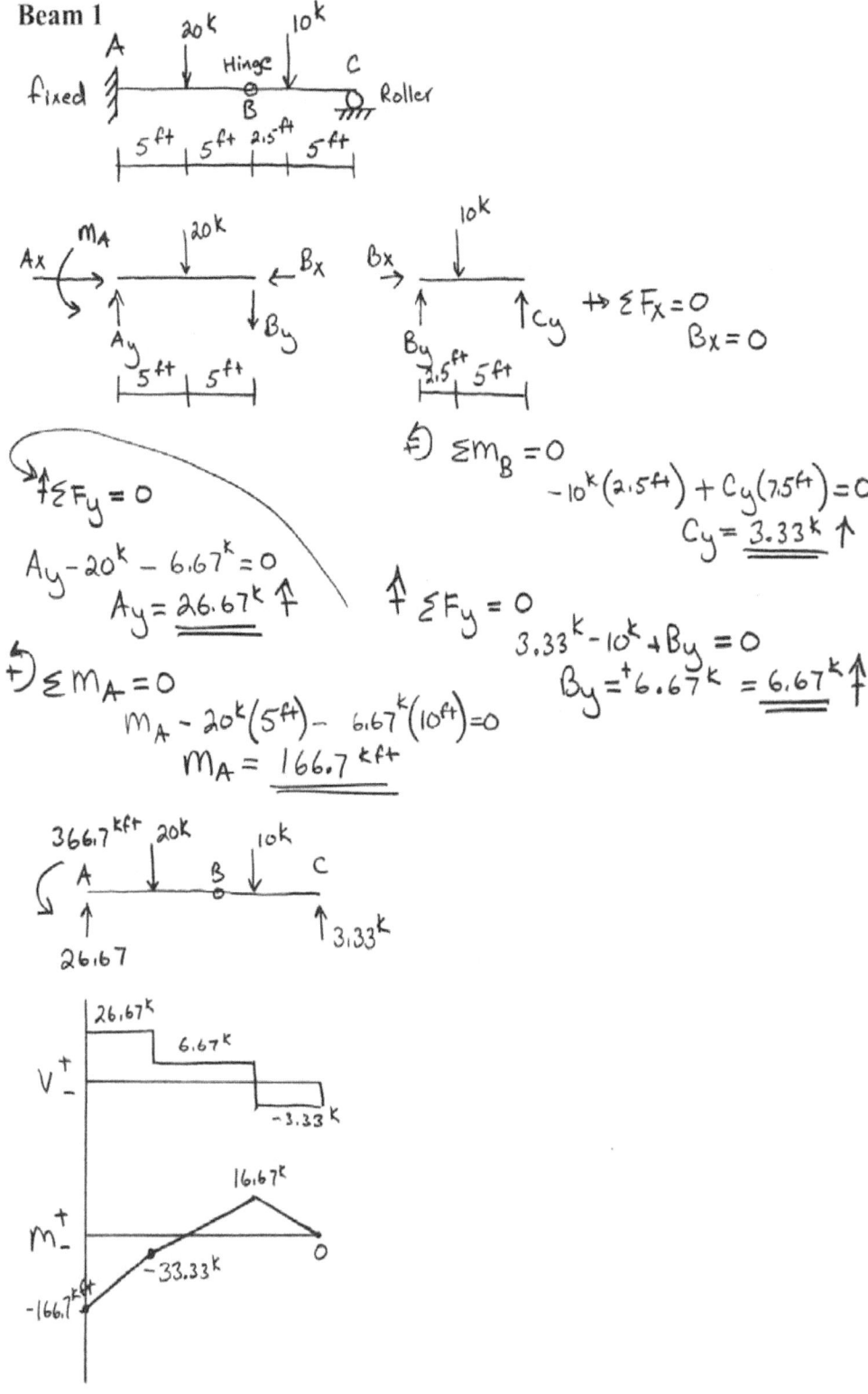

$$\leftrightarrow \Sigma F_x = 0$$
$$B_x = 0$$

$$\mathrel{\curvearrowleft} \Sigma m_B = 0$$
$$-10^k (2.5ft) + C_y (7.5ft) = 0$$
$$C_y = \underline{\underline{3.33^k}} \uparrow$$

$$+\uparrow \Sigma F_y = 0$$
$$A_y - 20^k - 6.67^k = 0$$
$$A_y = \underline{\underline{26.67^k}} \uparrow$$

$$\uparrow \Sigma F_y = 0$$
$$3.33^k - 10^k + B_y = 0$$
$$B_y = ^+6.67^k = \underline{\underline{6.67^k}}\uparrow$$

$$\mathrel{\curvearrowright} \Sigma m_A = 0$$
$$m_A - 20^k(5ft) - 6.67^k(10ft) = 0$$
$$m_A = \underline{\underline{166.7 \; kft}}$$

Beam 2

6^k A $2^{k/ft}$ 16^k B

Roller Pin

4 ft   14 ft   6 ft

$\rightarrow \Sigma F_x = 0$

$D_x = 0$

$+\circlearrowleft \Sigma m_A = 0$

$6^k(4ft) + 2^{k/ft}(4ft)(2ft) - 2^{k/ft}(20ft)(10ft)$
$\qquad -16^k(14ft) + B_y(20ft) = 0$

$\qquad\qquad B_y = \underline{\underline{29.2^k}} \uparrow$

$\uparrow \Sigma F_y = 0$

$\qquad -6^k - 2^{k/ft}(24ft) - 16^k + 29.2^k + A_y = 0$

$\qquad\qquad A_y = \underline{\underline{40.8^k}}$

$6K$ 16^k

$\rightarrow 0$

$\uparrow 29.2^k$

40.8^k

26.8^k

$V \begin{smallmatrix}+\\-\end{smallmatrix}$

-6^k $\leftarrow$ X $\rightarrow$

-14^k -1.2^k -17.2^k

139.56^{kft} -29.2^k

$m \begin{smallmatrix}+\\-\end{smallmatrix}$

$-40kft$

$26.8 - 2x = 0$

$X = \dfrac{26.8}{2}$

$X = \underline{\underline{13.4\,ft}}$

Beam 3

$10^k/ft$

A

$18 ft$ $16 ft$

$10^k/ft$

$A_x = 0$

A_y $B_y =$

$18 ft$ $16 ft$

83.33^k

$2°$ curve

A_1

V^+

A_2

$\leftarrow X \rightarrow$

$2°$ curve -86.67^k

$\pm 963 k ft$

m^+

$\overset{\curvearrowleft}{+} \, \Sigma M_A = 0$

$$B_y(34 ft) - \frac{1}{2}(10^k/ft)(18 ft)(\tfrac{2}{3})(18 ft)$$

$$-\frac{1}{2}(10^k/ft)(16 ft)(18 ft + 16\tfrac{ft}{3}) = 0$$

$$B_y = \underline{\underline{86.67^k}}$$

$\uparrow \Sigma F_y = 0$

$$-\frac{1}{2}(10^k/ft)(18 ft) - \frac{1}{2}(10^k/ft)(16 ft)$$

$$+ 86.67^k + A_y = 0$$

$$A_y = \underline{\underline{83.33^k}}$$

$$83.33^k - \frac{1}{2}(\tfrac{10}{18}x)(x) = 0$$

$$x = \underline{17.32^{ft}}$$

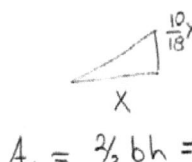

$$A_1 = \tfrac{2}{3}bh = \tfrac{2}{3}(17.32^{ft})(83.33^k)$$

$$A_1 = \underline{\underline{962.2 \ k ft}}$$

$$A_2 = \tfrac{2}{3}bh = \tfrac{2}{3}(16.68 \, ft)(-86.67)$$

$$A_2 = \underline{\underline{963 \ k ft}}$$

(rounding error)

Beam 4

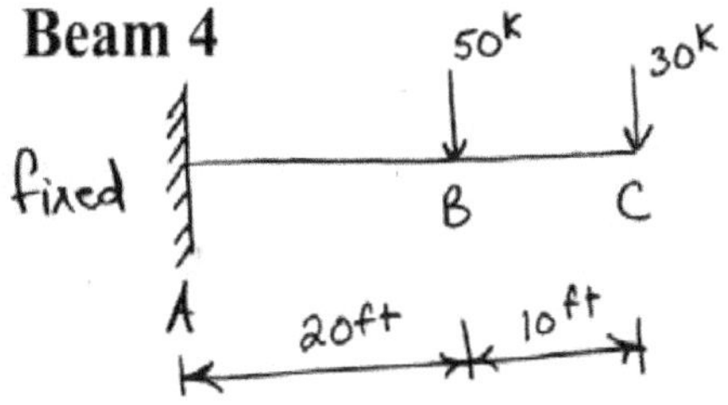

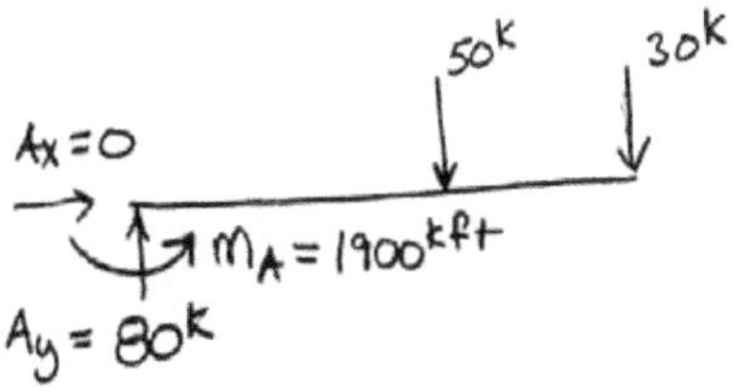

fixed

50^k 30^k

B   C

A

$20ft$ $10ft$

50^k 30^k

$A_x = 0$

$\curvearrowright M_A = 1900^{kft}$

$A_y = 80^k$

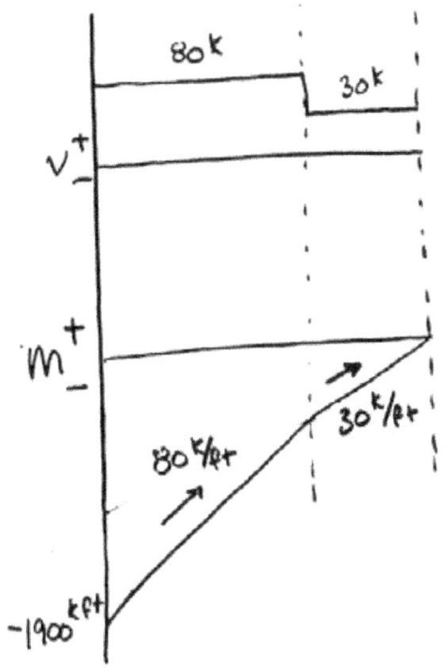

80^k

30^k

v^+_-

m^+_-

$80^k/ft$

$30^k/ft$

-1900^{kft}

$\uparrow \; \varepsilon F_y = 0$

$\quad -50^k - 30^k + A_y = 0$

$\qquad\qquad A_y = \underline{\underline{80^k}} \uparrow$

$\rightarrow \; \varepsilon F_x = 0 \quad A_x = \underline{\underline{0}}$

$\curvearrowleft \; \varepsilon M_A = 0$

$\quad M_A - 50^k(20ft) - 30^k(30ft) = 0$

$\qquad\qquad M_A = \underline{\underline{1900^{kft}}}$

Beam 5

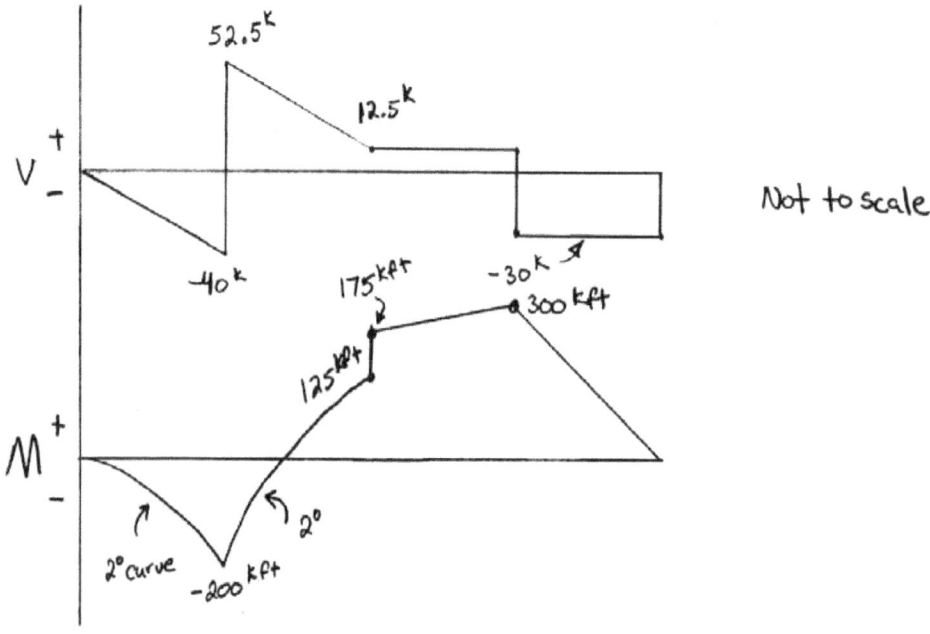

4 k/ft   C 50 kft   30k

A   B Pin   D   Roller   E

10 ft   10 ft   10 ft   10 ft

Draw the shear + moment diagrams for the beam shown.

$(+) \; \Sigma M_B = 0$

$4^{k/ft}(10^{ft})(5^{ft}) - 4^{k/ft}(10^{ft})(5^{ft}) - 50^{kft} + D_y(20^{ft}) + 30^k(30^{ft}) = 0$

$$D_y = \underline{\underline{42.5^k}} \downarrow$$

$\uparrow \Sigma F_y = 0$

$30^k - 4^{k/ft}(20^{ft}) - 42.5^k + A_y = 0$

$$A_y = \underline{\underline{92.5^k}} \uparrow$$

V⁺/⁻

52.5ᵏ

12.5ᵏ

-40ᵏ

M⁺/⁻

175 kft

125 kft

-30ᵏ

300 kft

2° curve   -200 kft   2°

Not to scale

Frames

Frame 1

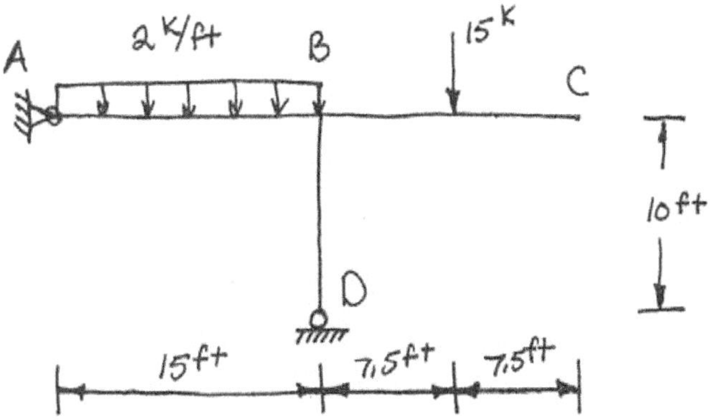

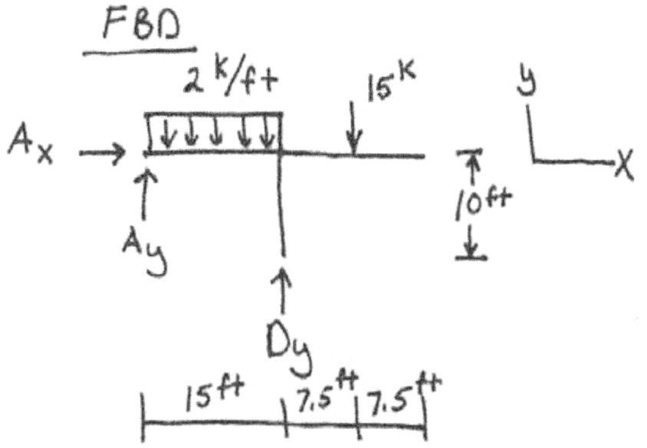

$$\circlearrowleft + \quad \Sigma M_A = 0$$

$$-2^{k/ft}(15^{ft})(7.5^{ft}) - 15^{k}(15^{ft} + 7.5^{ft}) + D_y(15^{ft}) = 0$$

$$D_y = \underline{\underline{37.5^{k}}}$$

$$+\uparrow \quad \Sigma F_y = 0 \qquad A_y - 2^{k/ft}(15^{ft}) - 15^{k} + 37.5^{k} = 0$$

$$A_y = \underline{\underline{7.5^{k}}}$$

$$\xrightarrow{+} \quad \Sigma F_x = 0 \qquad A_x = \underline{\underline{0}}$$

Frame 1 Break Frame into three members and solve for internal forces.

A   2 $^{k}/_{ft}$ B

ok →

M_{B_1}

7.5^{k} B_{y_1}

15 ft

↑ $\Sigma F_y = 0$
$$7.5^k - 2^{k/ft}(15^{ft}) + B_{y_1} = 0$$
$$B_{y_1} = 22.5^k$$

↻ $\Sigma m_A = 0$
$$-2^{k/ft}(15^{ft})(7.5^{ft}) + 22.5^k(15^{ft}) - M_{B_1} = 0$$
$$M_{B_1} = 112.5^{kft}$$

Area 1   Area 2

7.5^{k}
V $^+_-$

—X—

-22.5^k

$\underline{X\ distance}$
$$7.5 - 2X = 0$$
$$X = 3.75^{ft}$$

14.0625^{kft}

m $^+_-$

-112.5^{kft}

$$Area\ 1 = \tfrac{1}{2}(7.5^k)(3.75^{ft}) = 14.0625^{kft}$$
$$Area\ 2 = \tfrac{1}{2}(22.5^k)(11.25^{ft}) = -126.562^{kft}$$

B   15k   C

ok ↻ →

m_{B_2} ↑B_{y_2}

↑ $\Sigma F_y = 0$
$$B_{y_2} - 15^k = 0$$
$$B_{y_2} = 15^k$$

15^{k}

V $^+_-$

↻ $\Sigma m_B = 0$
$$+ m_{B_2} - 15^k(7.5^{ft}) = 0$$
$$m_{B_2} = 112.5^{kft}$$

0^{kft}

m $^+_-$

-112.5^{kft}

Member BD
    No shear or moment diagram.

37.5^{k} ↓
B

D
   ↑ 37.5^{k}

Frame 2

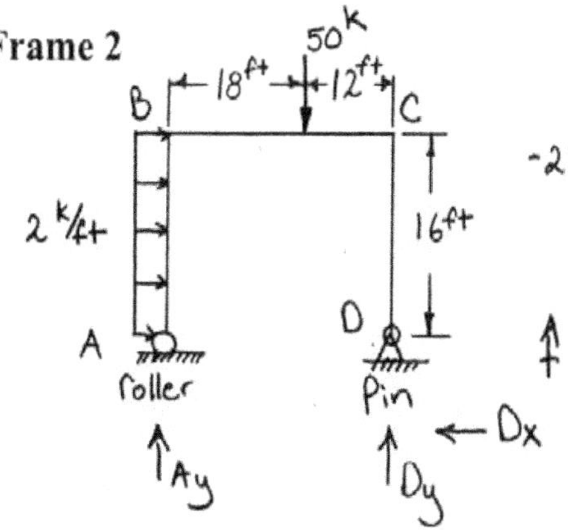

$$\overset{+}{\circlearrowleft)} \ \Sigma M_A = 0$$

$$-2^{k/ft}(16^{ft})(8^{ft}) - 50^k(18^{ft}) + D_y(30^{ft}) = 0$$

$$D_y = 38.53^k \uparrow$$

$$\uparrow \ \Sigma F_y = 0$$
$$A_y - 50^k + 38.53^k = 0$$
$$A_y = 11.47^k \uparrow$$

$$\rightarrow \ \Sigma F_x = 0$$
$$-D_x + 2^{k/ft}(16^{ft}) = 0$$
$$D_x = 32^k \leftarrow$$

Each segment of the frame is in equilibrium and can be evaluated separately.

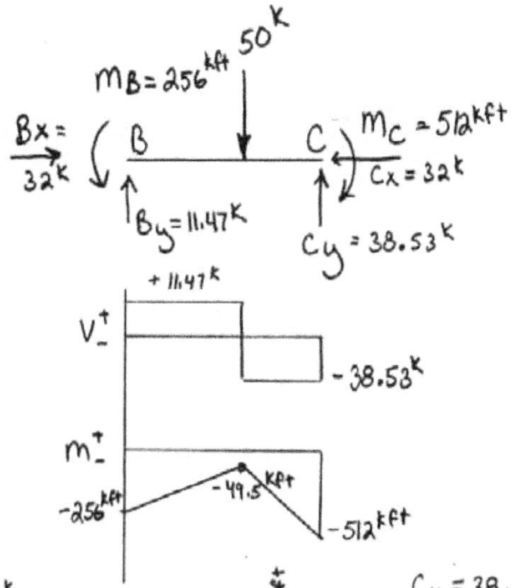

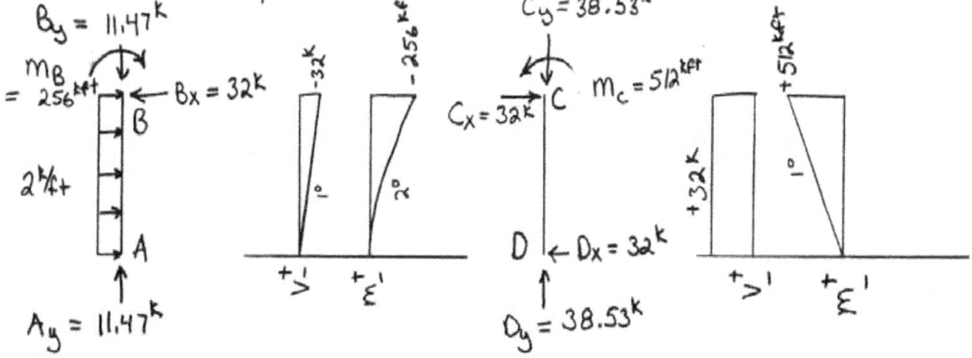

Frame 3

A 3^k/ft 10^k

B (Hinge)

18.75 ft

8^k

20 ft

C

15 ft   5 ft 10 ft   15 ft

Shear and moment diagrams.

FBD ②

$\circlearrowright$ $\Sigma M_A = 0$
$$-10^k(20ft) - 3^{k/ft}(15ft)(7.5ft) + B_y(30ft) = 0$$
$$B_y = 17.92^k$$

$\uparrow$ $\Sigma F_y = 0$
$$A_y + 17.92^k - 10^k - 3^{k/ft}(15ft) = 0$$
$$A_y = 37.08^k$$

A_x 3^k/ft 10^k

B

A_y FBD ①

5 / 3 / 4

8^k

C_x

C_y

FBD ②

A_x 3 k/ft 10^k

$\leftarrow B_x$

A_y $\llcorner_x^y$ B_y

FBD ①

$\uparrow$ $\Sigma F_y = 0$

$$37.08 - 3^{k/ft}(15ft) - 10^k - 8^k(\tfrac{3}{5}) + C_y = 0$$
$$C_y = 22.72^k$$
(whole frame)

37.08^k

37.08 - 3X = 0
X = 12.36 ft

V$_-^+$ $\leftarrow$X$\rightarrow$

-7.92^k

-17.92^k

229.2 k-ft

m$_-^+$

FBD ③

$\circlearrowright$ $\Sigma M_c = 0$
$$8^k(6.25ft) - B_y(25ft) = 0$$
$$B_y = 2^k$$

$\uparrow$ $\Sigma F_y = 0$
$$C_y - 8^k + 2^k = 0$$
$$C_y = 6^k$$

3x

2k

8^k

37.5 k-ft

-6k

Hint – Virtual Work

Always place the virtual unit load at the location and in the direction you think it will deflect.

ie)

Deflection at point C.

↑ use this to write "real" Moment functions for the beam (M)

+

↑ use this to write "virtual" moment functions for the beam (m_v)

korkn

$$1 * \Delta = \int \frac{m_v M}{EI} dx$$

Virtual moment functions (equations)
Real moment functions (equations)
moment of Inertia.
modulus of elasticity (usually steel $\Rightarrow E = 29,000 \, ^k/_{in^2}$)
integrate

Virtual unit load
Real Deflection

If you get an answer with a sign opposite to the one you were expecting, it means your virtual load is in the wrong direction.

Which means your deflection is opposite to what you originally thought.

Beam 1

Displacement at the Center of the beam (Point B) $EI = constant$

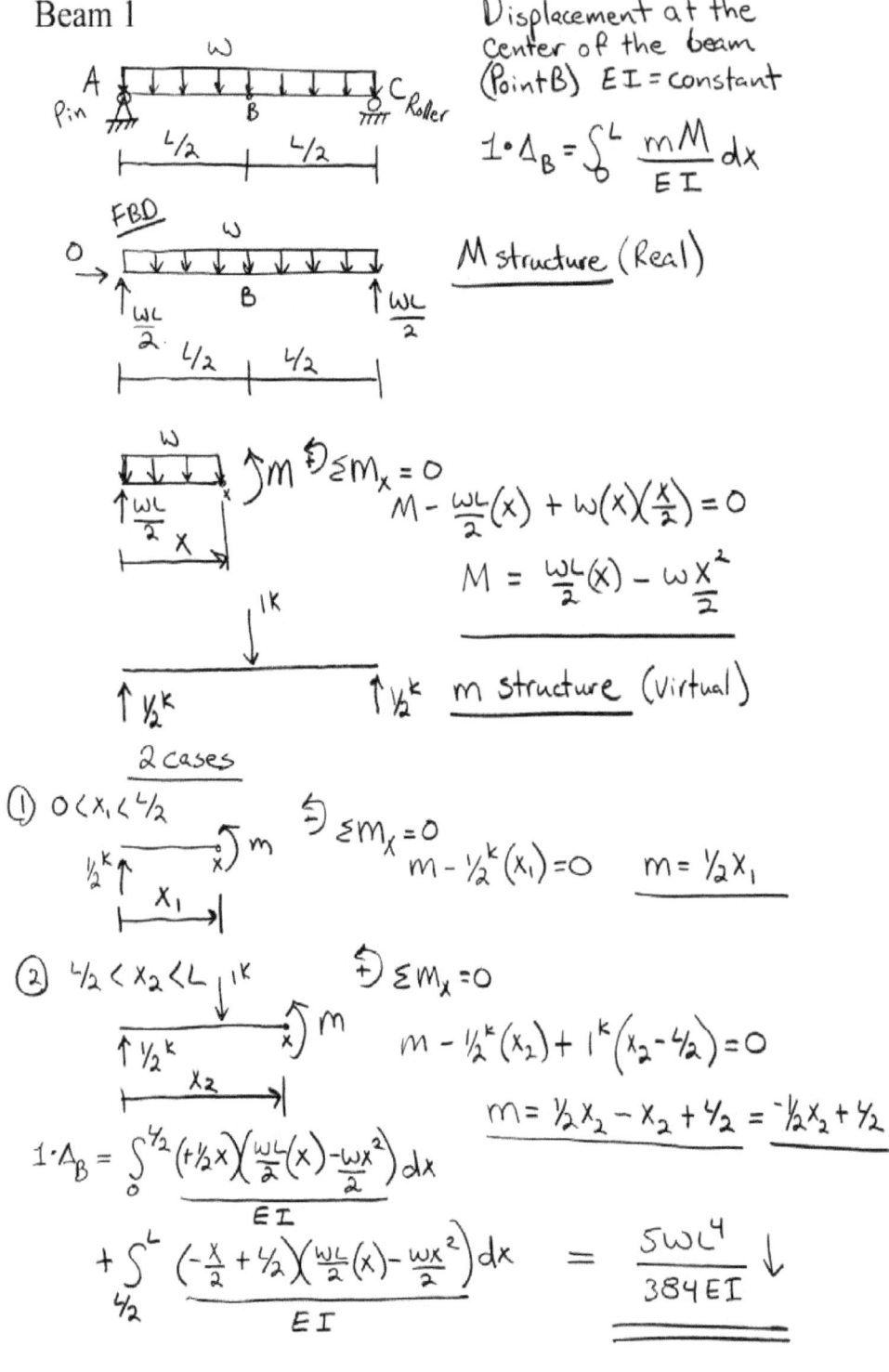

$$1 \cdot \Delta_B = \int_0^L \frac{mM}{EI} dx$$

$\underline{M \text{ structure (Real)}}$

FBD

$\sum M_x = 0$

$$M - \frac{WL}{2}(x) + W(x)\left(\frac{x}{2}\right) = 0$$

$$M = \frac{WL}{2}(x) - \frac{Wx^2}{2}$$

$\underline{m \text{ structure (Virtual)}}$

2 cases

① $0 < x_1 < \frac{L}{2}$

$\sum M_x = 0$

$$m - \frac{1}{2}^k (x_1) = 0 \qquad \underline{m = \frac{1}{2}x_1}$$

② $\frac{L}{2} < x_2 < L$

$\sum M_x = 0$

$$m - \frac{1}{2}^k (x_2) + 1^k \left(x_2 - \frac{L}{2}\right) = 0$$

$$\underline{m = \frac{1}{2}x_2 - x_2 + \frac{L}{2} = -\frac{1}{2}x_2 + \frac{L}{2}}$$

$$1 \cdot \Delta_B = \int_0^{L/2} \frac{\left(\frac{1}{2}x\right)\left(\frac{WL}{2}(x) - \frac{Wx^2}{2}\right)}{EI} dx$$

$$+ \int_{L/2}^L \frac{\left(-\frac{x}{2} + \frac{L}{2}\right)\left(\frac{WL}{2}(x) - \frac{Wx^2}{2}\right)}{EI} dx \qquad = \frac{5WL^4}{384EI} \downarrow$$

Vertical displacement at point C.   Beam 2

$\downarrow 25^k$ $_B 10^{k}/ft$ $EI = constant$

A

Pin $\stackrel{A}{\underset{\text{/////}}{}}$ $\underset{\text{Roller}}{}$ C $1^k \Delta_c = \int \dfrac{M m_v}{EI} dx$

$\overbrace{12.5ft}$ $\overbrace{12.5ft}$ $\overbrace{20 ft}$

25^k Real System

M   Limits   Integrate

$A_x = 0$ $0 < X_1 < 12.5^{ft}$ $\int_0^{12.5}$

$A_y = 57.5^k$ $B_y = 417.5^k$ C $0 < X_2 < 25^{ft}$ $\int_{12.5}^{25}$

$\vdash X_1 \dashv$ $\vdash X_3 \dashv$ $0 < X_3 < 20^{ft}$ $\int_0^{20}$

$\vdash X_2 \dashv$

                Virtual system 1^k
                     m_v

$A_x = 0$ C

$\downarrow A_y = .8^k$ $\uparrow B_y = 1.8^k$

$\vdash X_1 \dashv$

$\vdash X_2 \dashv$ $\vdash X_3 \dashv$

$\underline{X_1}$ $\underline{X_3}$

$10^k/ft$ $M = 57.5X - \dfrac{10 X^2}{2}$ $10^k/ft$

$57.5^k \vdash X_1 \dashv$ M $\vdash X_3 \dashv$

                                                    $M = \dfrac{-10 X^2/2}$

$.8^k \vdash X_1 \dashv$ $m_v = -.8X$ $m_v \vdash X_3 \dashv$ 1^k

$\underline{X_2}$ 25^k $m_v = -X$

$\vdash 12.5ft \dashv 10^k/ft$ M $M = \dfrac{-10X^2}{2} + 32.5X + 312.5$

$57.5^k \vdash X_2 \dashv$

$\downarrow .8^k$ m_v $m_v = -.8X$

$\vdash X_2 \dashv$

Beam 2

$$1^k * \Delta_C = \left[\int_0^{12.5} (57.5x - 5x^2)(-.8x)\,dx \; + \; \int_{12.5}^{25} (-5x^2 + 32.5x + 3125)(-.8x)\,dx \right.$$

$$\left. + \; \int_0^{20} (-5x^2)(-x)\,dx \right] \frac{1}{EI}$$

$$= \frac{383593.75^{kft^3}}{EI}$$

$$\int_0^{12.5} \frac{(-46x^2 + 4x^3)\,dx}{EI} \;\Rightarrow\; \frac{\frac{-46x^3}{3} + \frac{4x^4}{4}\Big|_0^{12.5}}{EI} = \frac{-5533.85}{EI}$$

$$\int_{12.5}^{25} \frac{(4x^3 - 26x^2 - 250x)\,dx}{EI} \;\Rightarrow\; \frac{\frac{4x^4}{4} - \frac{26x^3}{3} - \frac{250x^2}{2}\Big|_{12.5}^{25}}{EI} = \frac{189127.6}{EI}$$

$$\int_0^{20} \frac{(5x^3)\,dx}{EI} \;\Rightarrow\; \frac{\frac{5x^4}{4}\Big|_0^{20}}{EI} = \frac{200000}{EI}$$

$$\Delta_C = \frac{383593.75 \; kft^3}{EI}$$

$$\Delta_C = \frac{383593.75^{kft^3}}{EI} \quad \downarrow$$

Beam 3

Determine the deflection
of the beam at point C.
$E = 29,000 ksi$
$I = 500 in^4$

30^k Real System

A   B

$\uparrow 15^k$ $\uparrow 15^k$
  X

Virtual System

1^k

A   B
         P+C
$\uparrow .25^k$ $\uparrow .75^k$

$0 < X < 10^{ft}$

$\uparrow 15^k$ X $M = 15X$

$0 < X < 10^{ft}$

$\uparrow .25^k$ X $M_v = .25X$

30^k

$10^{ft} < X < 15^{ft}$

$\uparrow 15^k$ X

$\circlearrowleft \Sigma m = 0$

$M + 30(X-10) - 15X = 0$

$\underline{m = -15X + 300}$

$15^{ft} < X < 20$

$\underline{m = -15X + 300}$

$10^{ft} < X < 15^{ft}$

$\uparrow .25^k$ X $M_v = .25X$

1^k M_v

$15^{ft} < X < 20^{ft}$

$\uparrow .25^k$ X

$\circlearrowleft \Sigma m_v = 0$

$M_v + 1(X-15) - .25X = 0$

$\underline{M_v = -.75X + 15}$

$1^k * \Delta_c = \int \dfrac{M m_v}{EI} dx$

$1^k * \Delta_c = \dfrac{1}{EI} \left[\int_0^{10} (15X)(.25X)dx + \int_{10}^{15}(-15x+300)(.25x)dx + \int_{15}^{20}(-15x+300)(-.75x+15)dx \right]$

$\Delta_c = \dfrac{3437.5 \, kft^3}{EI} = \dfrac{3437.5 \, kft^3 (12^3 in^3/ft^3)}{(29,000 \frac{k}{in^2})(500 in^4)} = \underline{.411 \, in} \downarrow$

Beam 4

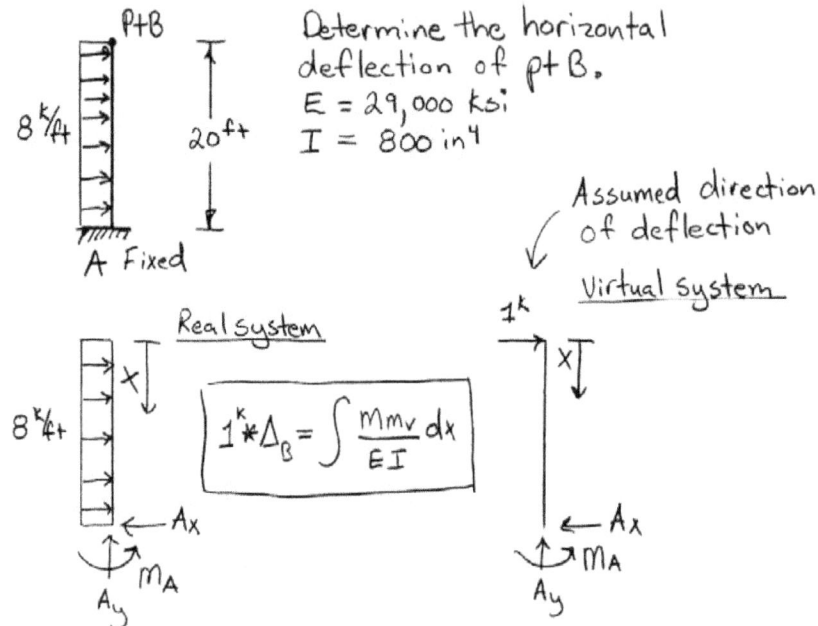

Determine the horizontal deflection of pt B.
$E = 29,000$ ksi
$I = 800$ in⁴

Assumed direction of deflection

Virtual System

Real System

$$1^k * \Delta_B = \int \frac{m m_v}{EI} dx$$

No need to find the reactions if your coordinate System starts at B.

$0 < x < 20$

$$m = \frac{8x^2}{2}$$

1^k

$0 < x < 20$

$$m_v = 1X$$

$$1^k * \Delta_B = \frac{1}{EI} \left[\int_0^{20} (4x^2)(X)dx \right] = \frac{4x^4}{4} \Big|_0^{20} = \frac{160,000 \, k^2 ft^3}{EI}$$

$$\Delta_B = \frac{160,000^{kft^3} (12^3 in^3 / ft^3)}{(29000 \frac{k}{in^2})(800 in^4)} = 11.9 \text{ in} \rightarrow$$

Truss 1

a) Find vertical displacement of Joint C.

b) Find the horizontal displacement of joint C.

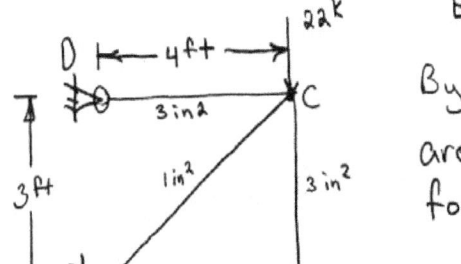

$E = 29,000^{ksi}$

By observation $AB + BC$ are zero force members for both deflections.

a) Need F_{DC} + F_{AC}

method of joints

Real system

F_{CD} ← 22^k ↓

F_{CA} $5\overbrace{}3$ over 4

$\uparrow \Sigma F_y = 0$

$-22^k + F_{CA}\left(\frac{3}{5}\right) = 0$

$F_{CA} = 36.67^k \,(C)$

$\rightarrow \Sigma F_x = 0$

$-F_{CD} + 36.67^k\left(\frac{4}{5}\right) = 0$

$F_{CD} = 29.34^k \,(T)$

for Vertical deflection

Virtual system

$D \leftarrow 4ft \rightarrow \downarrow 1^k$

$3ft$ ... A ... C

$F_{CA} = \frac{1}{22}\left(36.67^k\right) = 1.67^k \,(C)$

$F_{CD} = \frac{1}{22}\left(29.34^k\right) = 1.33^k \,(T)$

For horizontal deflection
Virtual system

$$\uparrow \Sigma F_y = 0$$
$$- F_{CA}\left(\tfrac{3}{5}\right) = 0$$
$$F_{CA} = 0$$

$$\Sigma F_x = 0$$
$$- F_{CD} + 1^k = 0$$
$$F_{CD} = \underline{\underline{1^k \ (T)}}$$

Member	Area (A)	Length (L)	Real Force (N)	Virtual Force n_v	Virtual Force n_h
DC	3in^2	48 in	29.34^k	1.33^k	1^k
CA	1in^2	60 in	-36.67^k	-1.67^k	0

a) Vertical displacement

$$\Delta_{cv} = \Sigma \frac{NnL}{AE} = \frac{(29.34^k)(1.33^k)(48)^{in}}{(3 \text{ in}^2)(29,000 \ ^k/_{in^2})} + \frac{(-36.67^k)(-1.67^k)(60 \text{ in})}{(1 \text{ in}^2)(29,000 \ ^k/_{in^2})}$$

$$\Delta_{cv} = \underline{\underline{.148^{in}}} \downarrow$$

b) horizontal displacement

$$\Delta_{ch} = \Sigma \frac{NnL}{AE} = \frac{(29.34^k)(1.00^k)(48 \text{ in})}{(3 \text{ in}^2)(29,000 \ ^k/_{in^2})} + \frac{(-36.67^k)(0)(60 \text{ in})}{(1 \text{ in}^2)(29,000 \ ^k/_{in^2})}$$

$$\Delta_{ch} = \underline{\underline{.016 \text{ in}}} \rightarrow$$

Truss 2

Determine the vertical deflection at joint C. E = 29,000 ksi

← Real System "N" values

$$\Delta_c = \sum \frac{NnL}{AE}$$

Joint A

$$\uparrow \, \Sigma F_y = 0$$

$$F_{AB}\left(\frac{1}{\sqrt{5}}\right) - 10^k = 0$$

$$F_{AB} = 22.36^k \; (T)$$

$$\rightarrow \Sigma F_x = 0$$

$$-F_{AD} + F_{AB}\left(\frac{2}{\sqrt{5}}\right) = 0$$

$$F_{AD} = 22.36\left(\frac{2}{\sqrt{5}}\right) = 20^k \, (C)$$

Joint B

$$\uparrow \, \Sigma F_y = 0$$

2 unknowns $\quad F_{BC}\left(\frac{5}{\sqrt{125}}\right) - 22.36^k\left(\frac{5}{\sqrt{125}}\right) - F_{BD} = 0$

$$\rightarrow \Sigma F_x = 0$$

$$F_{BC}\left(\frac{10}{\sqrt{125}}\right) - 22.36\left(\frac{10}{\sqrt{125}}\right) = 0$$

$$F_{BC} = 22.36^k \, (T)$$

$$F_{BD} = 0$$

$F_{CB} = 22.36^k \ (T)$

$\rightarrow \Sigma F_x = 0$

$F_{CO}\left(\dfrac{10}{\sqrt{200}}\right) - 22.36^k\left(\dfrac{10}{\sqrt{325}}\right) = 0$

$F_{CO} = 57^k \ (c)$

Virtual system "n" values.

Note: Since the 1^k load is at the same location as the 20^k load, the n values are $1/20$ of the N values.

Chart (tension is positive)

Member	N(k)	n(k)	L(in)	A(in)²	$\dfrac{NnL}{AE}$ (in)
AB	22.36	22.36/20	268.3	10	.023
BC	22.36	22.36/20	134.2	8	.014
BD	0	0/20	120	6	0
CD	⁻57	-57/20	169.7	8	+.12
AD	-20	-20/20	240	8	+.02

Total Σ +.18

$\Delta_C = .18 \ in \ \downarrow$

Truss 3

Determine the vertical deflection at pin D.
$E = 29,000$ ksi
$A = 5$ in^2

$\rightarrow \Sigma F_x = 0$

$50^k - A_x = 0$

$A_x = \underline{50}^k$

$\overset{+}{\curvearrowleft} \Sigma M_A = 0$

$-50^k(8^{ft}) + C_y(12^{ft}) = 0$

$C_y = \underline{33.33}^k$

$\uparrow \Sigma F_y = 0$

$33.33^k - A_y = 0 \quad A_y = \underline{33.33}^k \downarrow$

Joint A

$\overset{5}{\underset{3}{\triangle}}4$

$\uparrow \Sigma F_y = 0$

$F_{AB}(4/5) - 33.33^k = 0 \quad F_{AB} = \underline{41.66}^k \, (T)$

$\rightarrow \Sigma F_x = 0$

$F_{AD} + 41.66^k(3/5) - 50^k = 0$

$F_{AD} = \underline{25}^k \, (T)$

Joint D

$\uparrow \Sigma F_y = 0$

$F_{DB} = \underline{0}$

$\rightarrow \Sigma F_x = 0$

$F_{DC} - 25^k = 0 \quad F_{DC} = \underline{25}^k \, (T)$

Truss 3

Joint C

$$\uparrow \Sigma F_y = 0$$
$$- F_{CB} \left(\tfrac{4}{5}\right) + 33.33^k = 0$$
$$F_{CB} = \underline{41.66^k \ (C)}$$

Virtual System

Joint A

$$\uparrow \Sigma F_y = 0$$
$$-F_{AB} \left(\tfrac{4}{5}\right) + \tfrac{1}{2}^k = 0$$
$$F_{AB} = \underline{.625^k \ (C)}$$
$$\rightarrow \Sigma F_x = 0 \quad F_{CB} = \underline{.625^k \ (C)}$$
$$-.625\left(\tfrac{3}{5}\right) + F_{AD} = 0$$
$$F_{AD} = \underline{.375^k \ (T)}$$

Joint D

$$\uparrow \Sigma F_y = 0$$
$$F_{DB} = 1^k \ (T)$$

$$\rightarrow \Sigma F_x = 0 \quad F_{DC} = \underline{.375^k \ (T)}$$

$$1^k * \Delta_D = \Sigma \frac{N n L}{A E} \quad (T \text{ is positive})$$

member	$N_{(k)}$	$n_{(k)}$	L (in)	$\frac{N n L}{A E}$ (in)
A B	41.66	-.625	120	-.022
A D	25	.375	72	.005
B C	-41.66	-.625	120	.022
B D	0	1^k	96	0
D C	25	.375	72	.005
				$\Sigma \ .01''$

$$\Delta_D = \underline{.01''} \downarrow$$

Frame 1

B, 25ft, A, Real M, 10^k, C, $\leftarrow 20ft \rightarrow$

Find the deflection at point C in the vertical direction.

$E = 29,000\,ksi$ $\Big\}$ for all members
$I = 1000\,in^4$

Virtual m_v — B, C, 1^k, A

Real, 10^k

1^k

$200\,kft$, 10^k (Real)

$20\,kft$, $1K$ (Virtual)

200^{kft}, 10^k, x_2, 10^k

$m = -10x_2$
$0 < x_2 < 20$

$20\,kft$, $1k$, x_2, 1^k

$m_v = -x_2$

$200\,kft$, $10\,kft$, $200\,kft$, 10^k

$m = -200^{kft}$
$0 < x_1 < 25$

$20\,kft$, $20\,kft$, $1k$, 1^k

$m_v = -20^{kft}$

$$1 * \Delta_c = \int \frac{m_v m}{EI}\,dx$$

$$= \frac{1}{EI}\left[\int_0^{25}(-20)(-200)\,dx + \int_0^{20}(-x)(-10x)\,dx \right]$$

$$1 * \Delta_c = \frac{126,666.67\,k^2ft^3}{EI} = \frac{126,666.67\,k^2ft^3\left(12^2\,in^2/ft^2\right)}{\left(29,000\,k/in^2\right)\left(1000\,in^4\right)} = 7.5\,\frac{k\cdot in}{}\downarrow$$

$$\Delta_c = \underline{7.5\,in}\,\downarrow$$

Frame 2

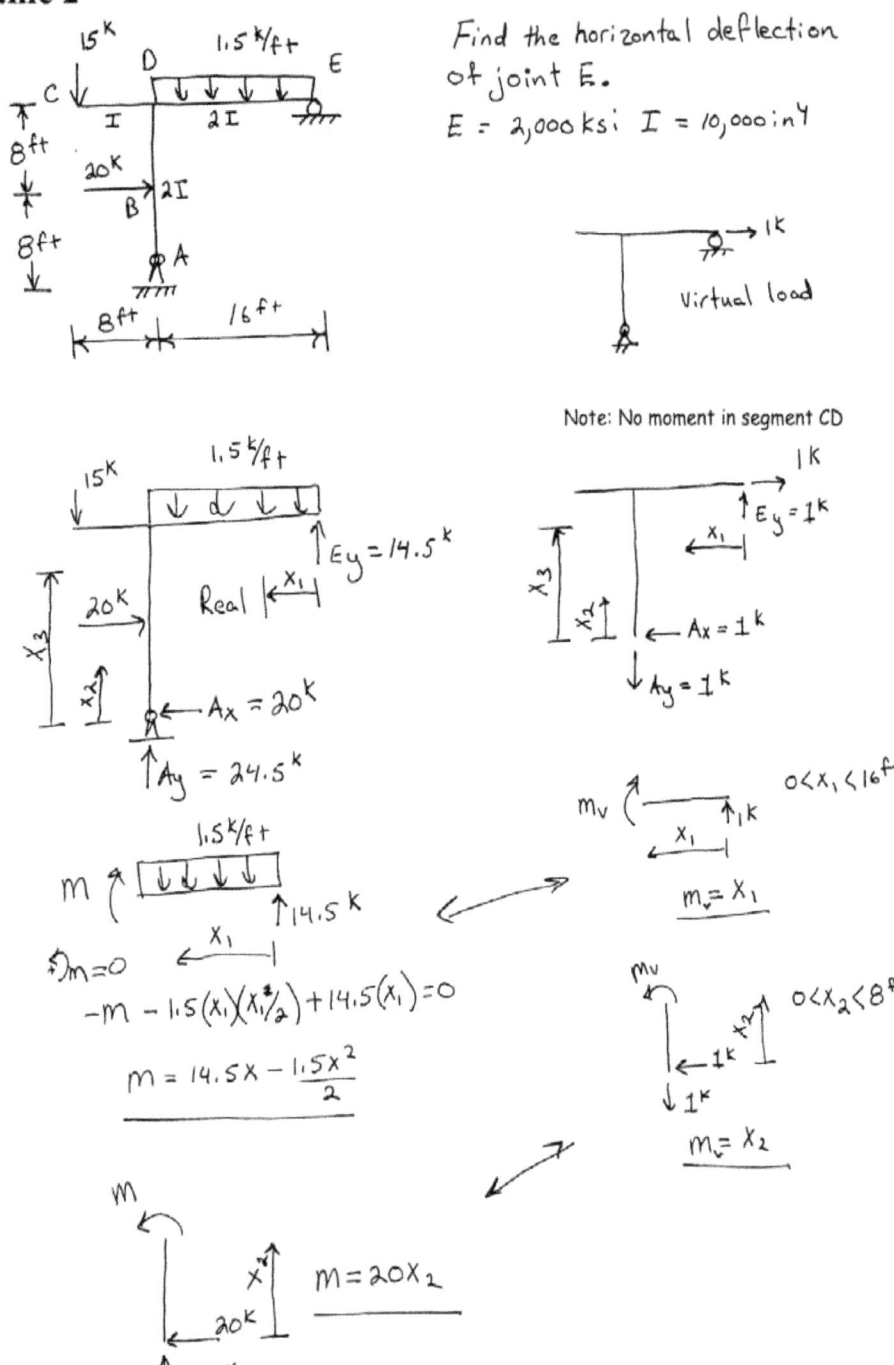

15^k

$1.5^{k}/ft$

C · D · E

I · 2I

8ft

20^k · B · 2I

8ft · A

8ft · 16ft

Find the horizontal deflection of joint E.

$E = 2,000$ ksi $\quad I = 10,000$ in⁴

→ 1^k

Virtual load

Note: No moment in segment CD

15^k · $1.5^{k}/ft$

Real · ← X_1 → · $\uparrow E_y = 14.5^k$

20^k

X_3 · X_2

← $A_x = 20^k$

$\uparrow A_y = 24.5^k$

$1.5^{k}/ft$

$m \uparrow$ · $\uparrow 14.5^k$

← X_1

$\sum m = 0$

$-m - 1.5(X_1)(X_1/2) + 14.5(X_1) = 0$

$$m = 14.5X - \frac{1.5X^2}{2}$$

m

← 20^k · X_2

$\uparrow 24.5^k$

$$m = 20X_2$$

1^k

$\uparrow E_y = 1^k$ · ← X_1

X_3 · X_2 · ← $A_x = 1^k$

$\downarrow A_y = 1^k$

m_v ← · $\uparrow 1^k$

← X_1 · $0 < X_1 < 16$ ft

$$m_v = X_1$$

m_v

← 1^k · X_2

$\downarrow 1^k$ · $0 < X_2 < 8$ ft

$$m_v = X_2$$

Frame 2

$8^{ft} < X_3 < 16^{ft}$

m_v $m = X_3$

$\Sigma m = 0$

$$m + 20(x-8) - 20x = 0$$

$$m = 20x - 20x + 160$$

$$\underline{m = 160}$$

$$1^k * \Delta_E = \int \frac{m_v m}{EI} dx$$

$$= \frac{1}{EI}\left[\int_0^{16} \frac{(x)(14.5x - .75x^2)}{2} dx + \int_0^8 \frac{(x)(20x)}{2} dx \right.$$

$$\left. + \int_8^{16} \frac{(x)(160)}{2} dx \, dx \right] = \frac{13,141.333 \, k^2 ft^3}{EI}$$

$$\Delta_E = \frac{13141.333 \, k ft^3 (12^3 in^3/ft^3)}{(16,000 in^4)(2,000 ksi)} = \underline{1.14 in} \longrightarrow$$

Frame 3

Determine the horizontal displacement of roller B.

$E = 29,000 \, ksi$
$I = 400 \, in^4$ for each member

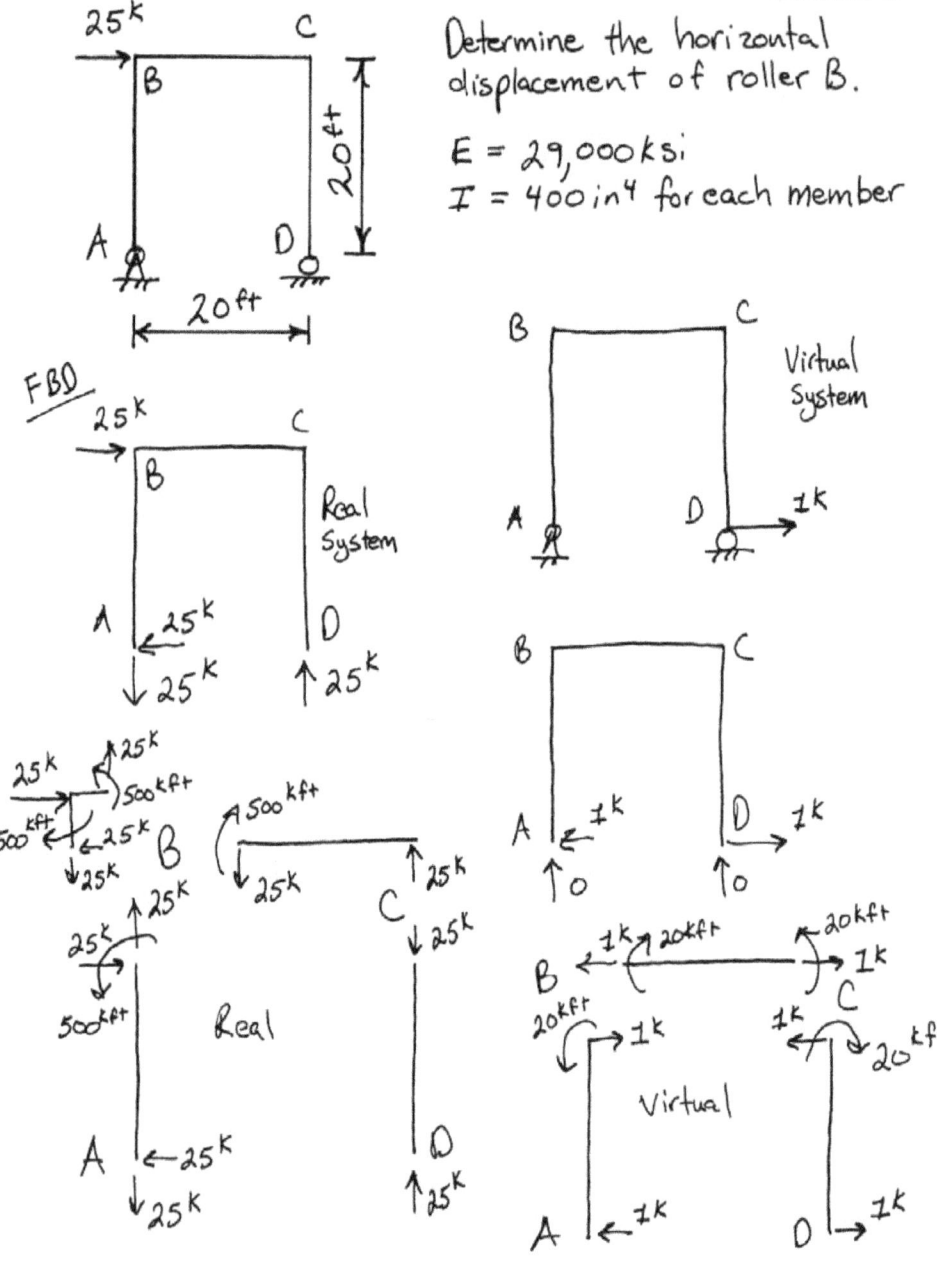

Coordinate System (for both Real + virtual)

Since member CD had no moment, no coordinate system is needed.

Real
$$0 < X_1 < 20^{ft}$$

$$m = 25X_1$$

$$0 < X_2 < 20^{ft}$$

$$m = 25X_2$$

Virtual
$$0 < X_1 < 20^{ft}$$

$$m_v = X_1$$

$$0 < X_2 < 20^{ft}$$

$$m_v = 20$$

$$1^* \Delta_D = \int \frac{m m_v}{EI} dx$$

$$1^k * \Delta_D = \frac{1}{EI} \left[\int_0^{20} (25X_1)(X_1) dx + \int_0^{20} (25X_2)(20) dx \right]$$

$$1^k * \Delta_D = \frac{166,666.67 \; ^{k^2 ft^3}}{EI}$$

$$= \frac{166,666.67 \; ^{k^2 ft^3}(12^3 in^3/ft^3)}{(29000 \frac{k}{in^2})(400 in^4)} = 24.8 \; ^{k in}$$

$$\Delta_D = 24.8 \, in \longrightarrow (excessive)$$

Hints - Visual Integration

$$1 * \Delta = \sum \frac{Ah}{EI}$$

↙ Area of real structure Moment diagram

↰ Value of the virtual moment diagram at the location of the centroid of the real moment diagram.

ie)

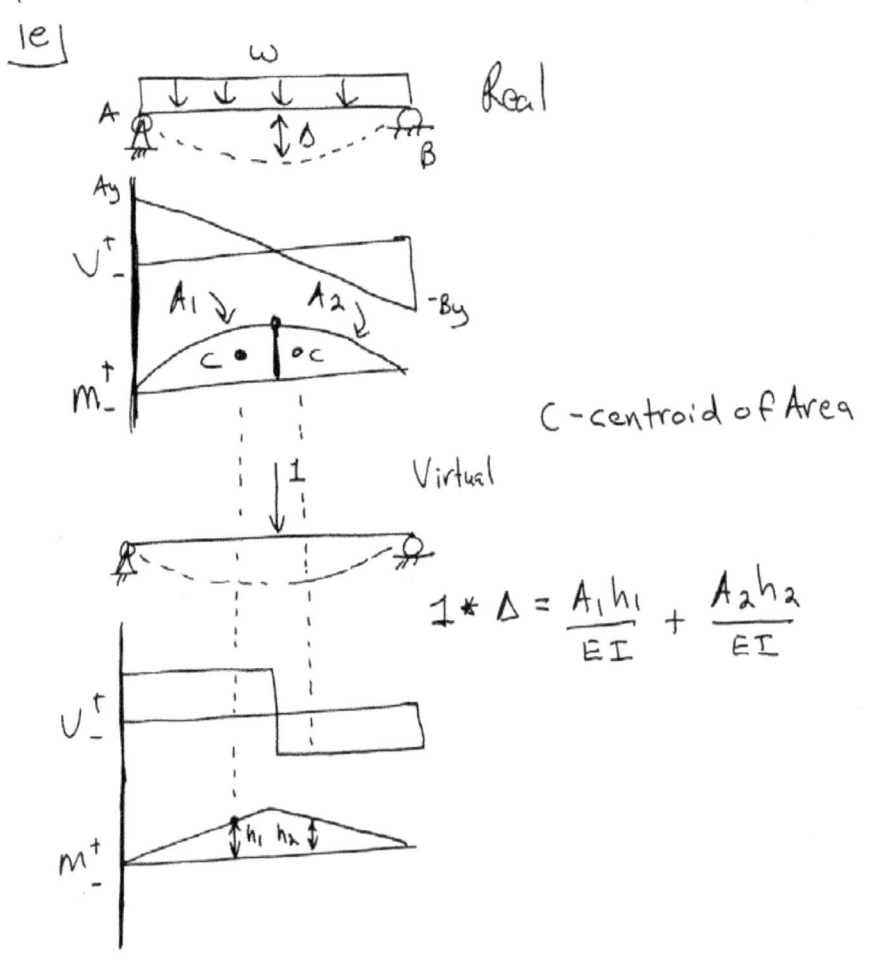

Real

C - centroid of Area

Virtual

$$1 * \Delta = \frac{A_1 h_1}{EI} + \frac{A_2 h_2}{EI}$$

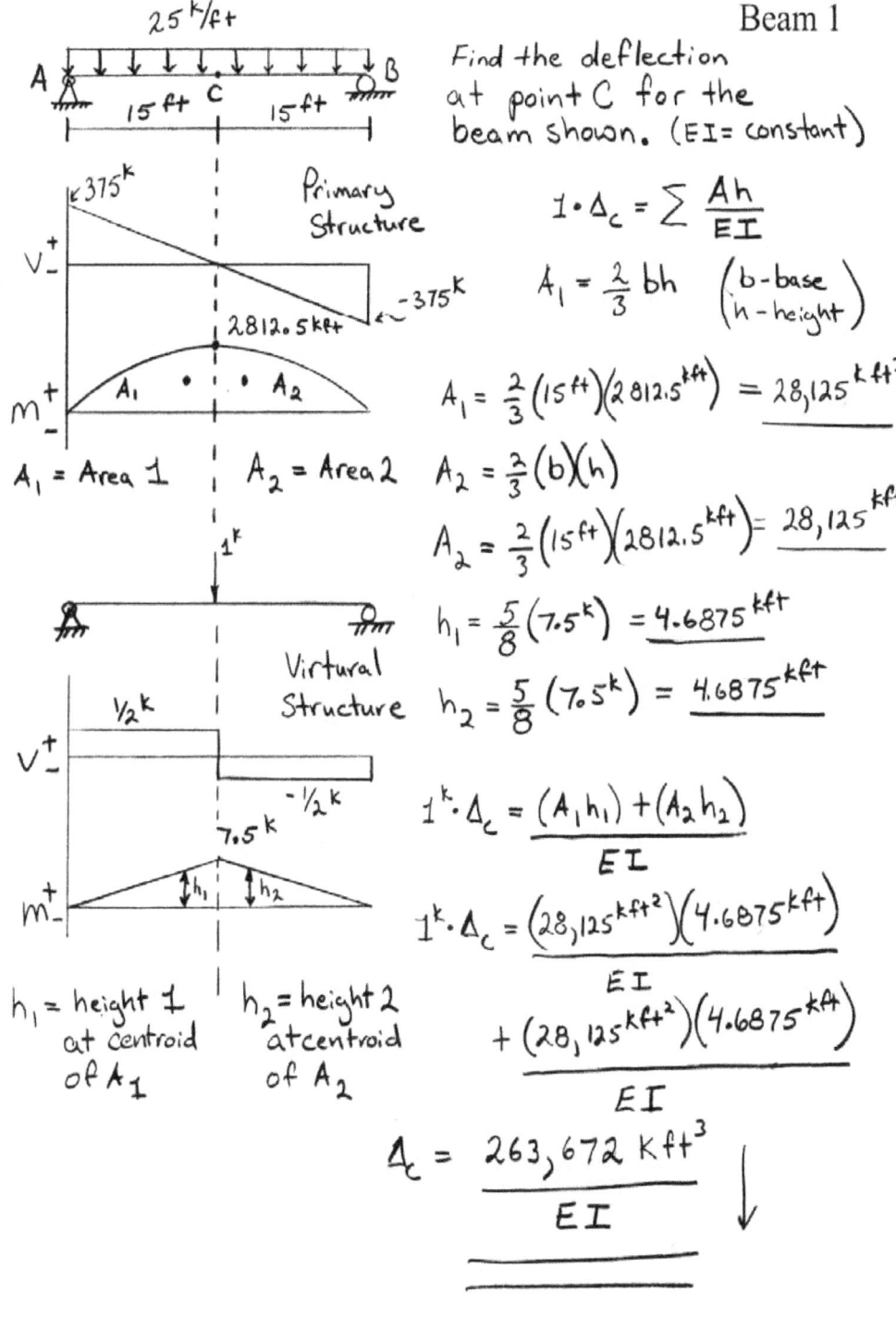

Beam 1

$25^{k}/ft$

A ↓↓↓↓↓↓↓↓↓↓ B

15 ft   C   15 ft

Find the deflection at point C for the beam shown. (EI = constant)

$$1 \cdot \Delta_c = \sum \frac{Ah}{EI}$$

Primary Structure

375^{k}

V_{-}^{+}

-375^{k}

$2812.5 \, kft$

m_{-}^{+} A_1 • A_2

$A_1 = $ Area 1 $A_2 = $ Area 2

$$A_1 = \frac{2}{3} bh \quad \left(\begin{array}{l} b\text{-base} \\ h\text{-height} \end{array}\right)$$

$$A_1 = \frac{2}{3}(15ft)(2812.5^{kft}) = 28,125^{kft^2}$$

$$A_2 = \frac{2}{3}(b)(h)$$

$$A_2 = \frac{2}{3}(15ft)(2812.5^{kft}) = 28,125^{kft}$$

Virtual Structure

1^{k}

$\frac{1}{2}^{k}$

V_{-}^{+}

$-\frac{1}{2}^{k}$

7.5^{k}

m_{-}^{+} $\downarrow h_1$ $\downarrow h_2$

$$h_1 = \frac{5}{8}(7.5^{k}) = \underline{4.6875 \, kft}$$

$$h_2 = \frac{5}{8}(7.5^{k}) = \underline{4.6875 \, kft}$$

$h_1 = $ height 1 at centroid of A_1

$h_2 = $ height 2 at centroid of A_2

$$1^{k} \cdot \Delta_c = \frac{(A_1 h_1) + (A_2 h_2)}{EI}$$

$$1^{k} \cdot \Delta_c = \frac{(28,125^{kft^2})(4.6875^{kft})}{EI}$$

$$+ \frac{(28,125^{kft^2})(4.6875^{kft})}{EI}$$

$$\Delta_c = \frac{263,672 \, Kft^3}{EI} \quad \downarrow$$

Beam 2

A   12 ^{kN}/m   B   Determine the displacement
fixed ↓↓↓↓↓↓↓↓   of point B.
                      E = 200 6PA
      10m   I = 500(10⁶) mm⁴

A   12 ^{kN}/m
M_A ↓↓↓↓↓↓↓↓   ↑ $\Sigma F_y = 0$ $A_y = 12 \frac{kN}{m}(10m)$
  ↑A_y Primary $A_y = 120 kN$
  120kN Structure
                  ↺ $\Sigma M_A = 0$ $M_A = 12 \frac{kN}{m}(10m)(5m)$
                                      $M_A = 600 kNm$ ↻

V^+_-

m^+_- x | $x = \frac{1}{4}b = \frac{1}{4}(10) = 2.5m$ $1^{kN} \cdot \Delta_B = \Sigma \frac{Ah}{EI}$
           * Centriod
         A   Area of Primary Structure
        -600 ^{kNm}
                        1kN $A = \frac{1}{3}bh = \frac{1}{3}(10m)(-600 kNm)$
      Virtual load   ↓
                        ↓^{1kN} $A = -2000 kNm^2$
-10kNm
  ↺ ↑₁kN   Value of virtual moment
                                    at $x = 2.5m$

                        1 kN $m = 1x - 10 = 1(2.5m) - 10m$
V^+_- $m = -7.5 kNm$

m^+_- ↕h $1^{kN} \cdot \Delta_B = \frac{Ah}{EI} = \frac{-2000 kNm^2(-7.5 kNm)}{200(10^6)\frac{kN}{m^2}(500(10^6)mm^4)}$
     -10 ^{kNm}

                                    $\Delta_B = .15m = 150 mm$ ↓

Beam 3

Determine the displacement of the beam at point B.
EI = constant

$$1^{k} \cdot \Delta_{B} = \sum \frac{Ah}{EI}$$

$$A_{1} = \frac{2}{3}bh = \frac{2}{3}\left(\frac{L}{2}\right)\left(\frac{WL^{2}}{4}\right)$$

$$A_{1} = \frac{WL^{3}}{12}$$

$$A_{2} = \frac{2}{3}bh = \frac{2}{3}\left(\frac{L}{2}\right)\left(\frac{WL^{2}}{4}\right)$$

$$A_{2} = \frac{WL^{3}}{12}$$

$$h_{1} = \frac{1}{2}\left(\frac{5}{8}\right)L = \frac{5}{16}L$$

$$h_{2} = \frac{1}{2}\left(\frac{5}{8}\right)L = \frac{5}{16}L$$

$$1^{k} \cdot \Delta_{C} = \frac{A_{1}h_{1}}{EI} + \frac{A_{2}h_{2}}{EI}$$

$$= \frac{WL^{3}}{12}\left(\frac{5}{16}\right)L + \frac{WL^{3}}{12}\left(\frac{5}{16}\right)L$$

$$= \frac{5WL^{4}}{384EI} \quad \downarrow$$

Beam 4

Determine the deflection at Point C.
$E = 29,000$ ksi
$I = 500$ in^4

Real System

Virtual System

$$1^k * \Delta_c = \sum \frac{Ah}{EI}$$

$A_1 = \frac{1}{2}(10^{ft})(150^{kft}) = \underline{750 \text{ Kft}^2}$ $h_1 = \frac{2}{3}(10^{ft})(.25) = \underline{1.67}^{kft}$

$A_3 = \frac{1}{2}(5^{ft})(75^{kft}) = \underline{187.5}^{kft^2}$ $h_1 = \frac{2}{3}(5^{ft})(.75) = \underline{2.5}^{kft}$

A_2

$A_2 = \frac{5^{ft}(150+75)}{2} = \underline{562.5}^{kft^2}$

$X = \frac{5^{ft}(2(75)+150)}{3(150+75)} = \underline{2.22}^{ft}$

$h_1 = (15^{ft} - 2.22)(.25) = \underline{3.19}^{kft}$

$$\Delta_c = \frac{\left((750)(1.67) + (187.5)(2.5) + (562.5)(3.19)\right)^{kft^3}\left(\frac{12^3 in^3}{ft^3}\right)}{\left(29000 \frac{k}{in^2}\right)(500 in^4)}$$

$$\Delta_c = \underline{.41 \text{ in}} \downarrow$$

Determine the vertical
deflection of point C.
E = 29,000 ksi
I = 1000 in⁴

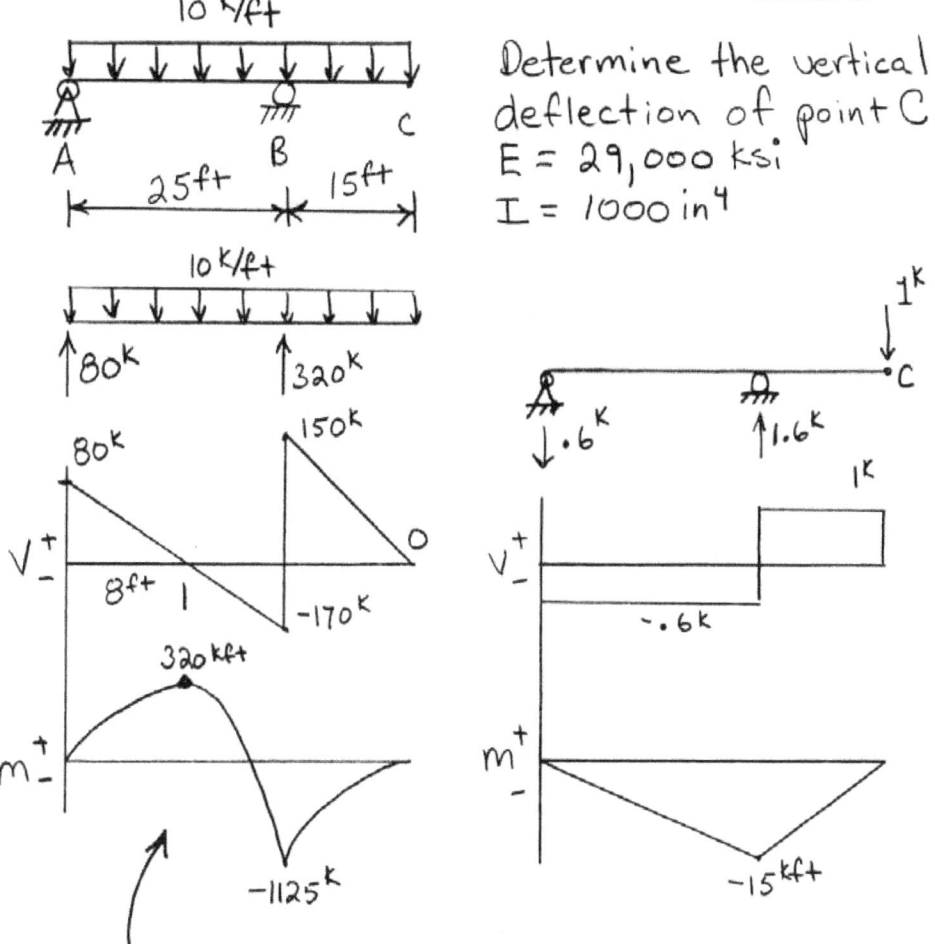

10 k/ft

A   25ft   B   15ft   C

10 k/ft

↑80ᵏ   ↑320ᵏ

80ᵏ   150ᵏ

V⁺₋   8ft   -170ᵏ   O

320 kft

m⁺₋   -1125ᵏ

1ᵏ ↓   •C

↓.6ᵏ   ↑1.6ᵏ

1ᵏ

V⁺₋   -.6ᵏ

m⁺₋   -15 kft

Since the real system has a moment
diagram that has difficult areas to
quantify try the "Cantilever beam parts"
method.

Place a fixed support at point B
and draw the moment diagram
for each individual force.

Beam 5

Real   Virtual

A $10^k/ft$ B $10^k/ft$ A   B $C_{\downarrow 1k}$

$\leftarrow 25^{ft} \rightarrow \leftarrow 15^{ft} \rightarrow$ C $\downarrow -.6^k$

                   $25 ft$ $15 ft$

V_-^+ 150^k $1k$

       -250^k V_-^+ $-.6^k$

         A_2

m_-^+ A_1 -1125^{kft} m_-^+ h_3 h_1 h_2

         -3125^{kft} -15^{kft}

B $h_1 = -11.25^{kft}$ @ $\frac{3}{4}(25^{ft})$

        C

$\uparrow 80^k$ $h_2 = -11.25^{kft}$ @ $\frac{3}{4}(15^{ft})$

    80^k $h_3 = -10^{kft}$ @ $\frac{2}{3}(25^{ft})$

V_-^+

m_-^+ A_3 2000^{kft}

$$1^k * \Delta_c = \sum \frac{Ah}{EI}$$

$$= \frac{1}{EI}\left[\frac{1}{3}(25^{ft})(-3125^{kft})(-11.25^{kft}) + \frac{1}{3}(15^{ft})(1125^{kft})(-11.25^{kft}) \right.$$

$$\left. + \frac{1}{2}(25^{ft})(2000^{kft})(-10^{kft}) \right]$$

$$= \frac{106250^{k^2ft^3}}{EI} = \frac{106250^{k^2ft^3}(12^3 in^3/ft^3)}{(29000 \frac{k}{in^2})(1000 in^4)}$$

$$\Delta_c = \underline{6.33 \, in \downarrow}$$

Frame 1

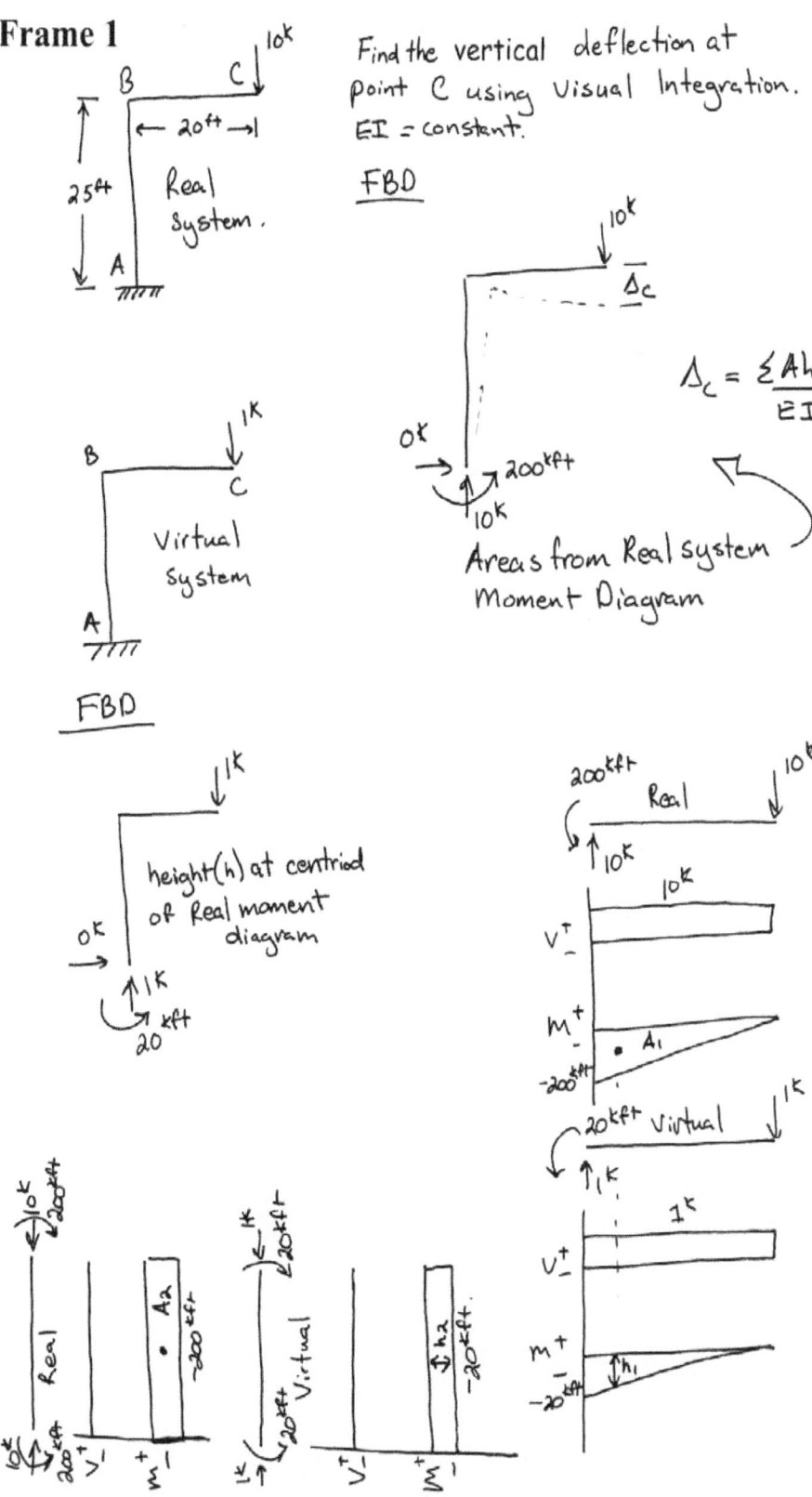

Find the vertical deflection at point C using Visual Integration. EI = constant.

$$\Delta_c = \sum \frac{Ah}{EI}$$

Areas from Real System Moment Diagram

Frame 1

$$1 \ast \Delta_c = \sum \frac{Ah}{EI} = \frac{A_1 h_1}{EI} + \frac{A_2 h_2}{EI}$$

$$\Delta_c = \frac{\frac{1}{2}\left(-200^{kft}\right)\left(20^{ft}\right)\left(-13.33^{kft}\right)}{EI}$$

$$+ \frac{\left(-200^{kft}\right)\left(25^{ft}\right)\left(-20^{kft}\right)}{EI} = \frac{126,660 \; k^2 ft^3}{EI}$$

If $E = 29,000 \; ksi$

$+ \quad I = 1000 \, in^4 \quad$ then

$$\Delta_c = \frac{126,660 \; kft^3 \left(\frac{12^3 in^3}{1\,ft^3}\right)}{\left(29,000 \, \frac{k}{in^2}\right)\left(1000 \, in^4\right)} = \underline{\underline{7.5''}} \downarrow$$

Frame 2

Find the horizontal deflection of point C.

$E = 29,000$ ksi

$I = 2000$ in^4

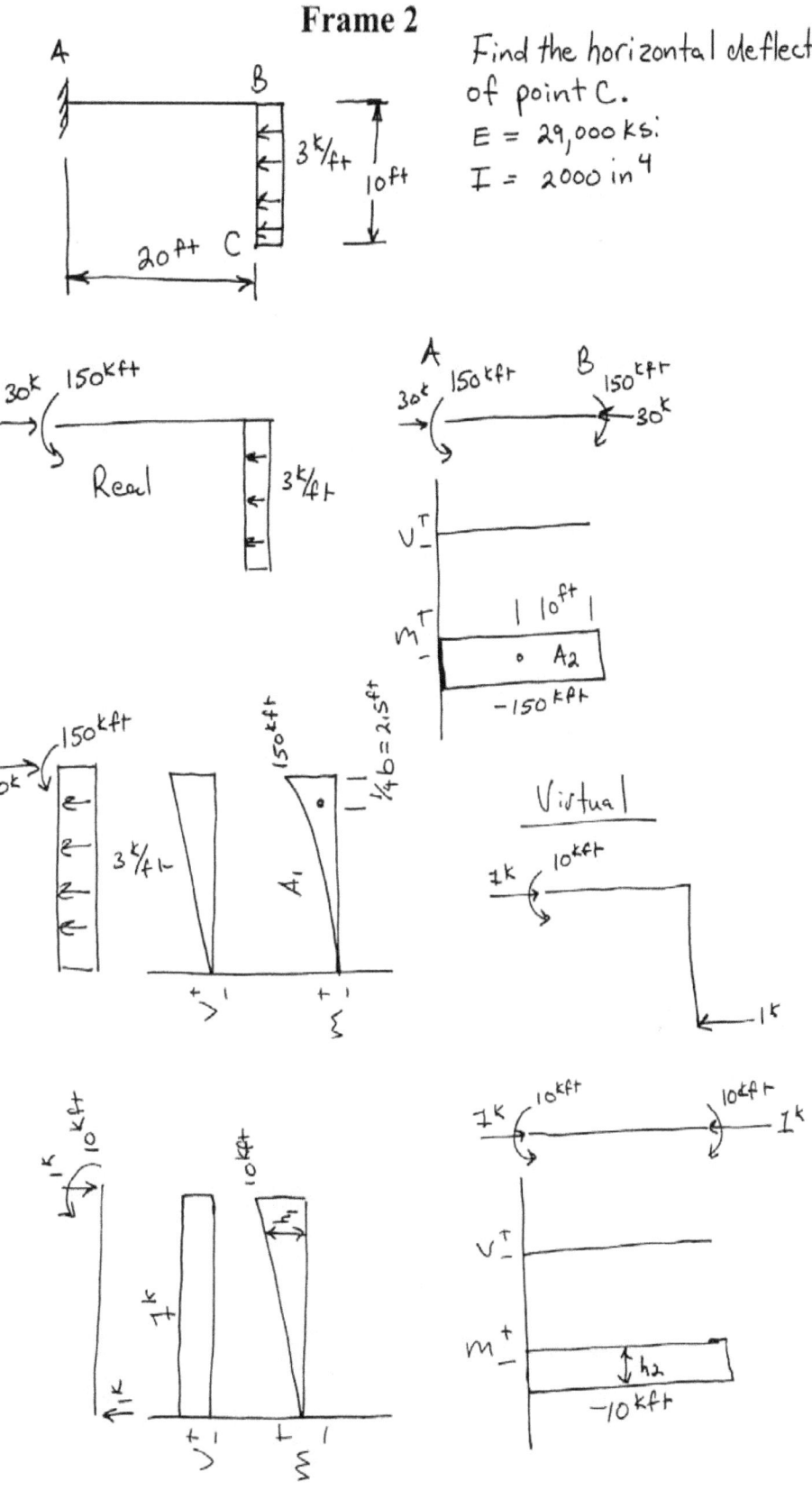

$$1^{k} \cdot \Delta_{char} = \frac{\sum Ah}{EI} = \frac{\frac{1}{3} bh(h_1)}{EI} + \frac{bh(h_2)}{EI}$$

$$= \frac{\frac{1}{3}(10^{ft})(150^{kft})(7.5^{kft})}{EI} + \frac{(20^{ft})(-150^{kft})(-10^{kft})}{EI}$$

$$= \frac{33,750^{kft^3}}{EI} = \frac{33,750 \, kft^3 \left(12in\right)^3 / ft^3}{\left(29,000 \frac{k}{in^2}\right)\left(2000 \, in^4\right)}$$

$$= \underline{\underline{1.0"}}$$

Determine the horizontal
displacement of roller B.

$E = 29,000\,ksi$
$I = 400\,in^4$ for each member

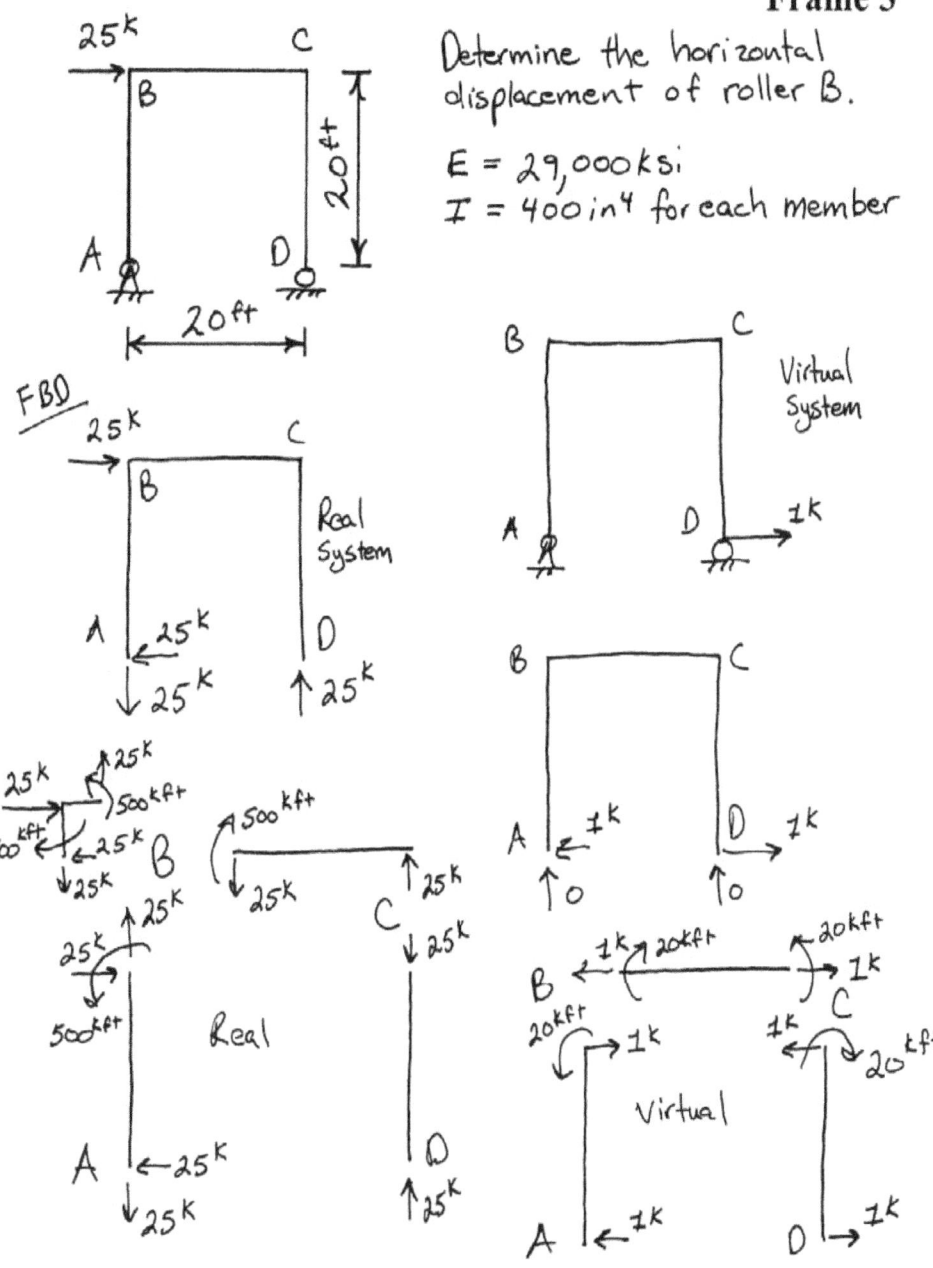

No moment diagram for member CD.   **Frame 3**

Real system (left): Column AB with $500^{k ft}$ moment, 25^k forces at B and A. Moment diagram A_1, $40/3$ ft, $500^{k ft}$.

Virtual system (right): Column AB with $20^{k ft}$ moment, 1^k forces. Moment diagram h_1, $40/3$ ft, $20^{k ft}$.

Member BC real: $B \to 500^{k ft}$, C, 25^k down, 25^k up.

V_-^+ diagram: -25^k

m_-^+ diagram: $500^{k ft}$, A_2

Member BC virtual: 1^k, B, $20^{k ft}$, $20^{k ft}$, 1^k, C

V_-^+ diagram

m_-^+ diagram: $20^{k ft}$, h_2

$$1^k * \Delta_B = \Sigma \frac{Ah}{EI}$$

$$1^k * \Delta_B = \frac{A_1 h_1}{EI} + \frac{A_2 h_2}{EI} = \frac{\frac{1}{2}\left(500^{k ft}\right)\left(20^{ft}\right)\left(13.33^{k ft}\right)}{EI}$$

$$+ \frac{\frac{1}{2}\left(500^{k ft}\right)\left(20^{ft}\right)\left(20^{k ft}\right)}{EI} = \frac{166,666.67}{EI}^{k^2 ft^3}$$

$$\Delta_B = \frac{166,666.67}{EI}^{k ft^3} = \frac{166,666.67^{k ft^3}\left(12^3 in^3 / ft^3\right)}{\left(29000 \frac{k}{in^2}\right)\left(400 in^4\right)} = \underline{\underline{24.8 in}} \rightarrow$$

Force Method Hints

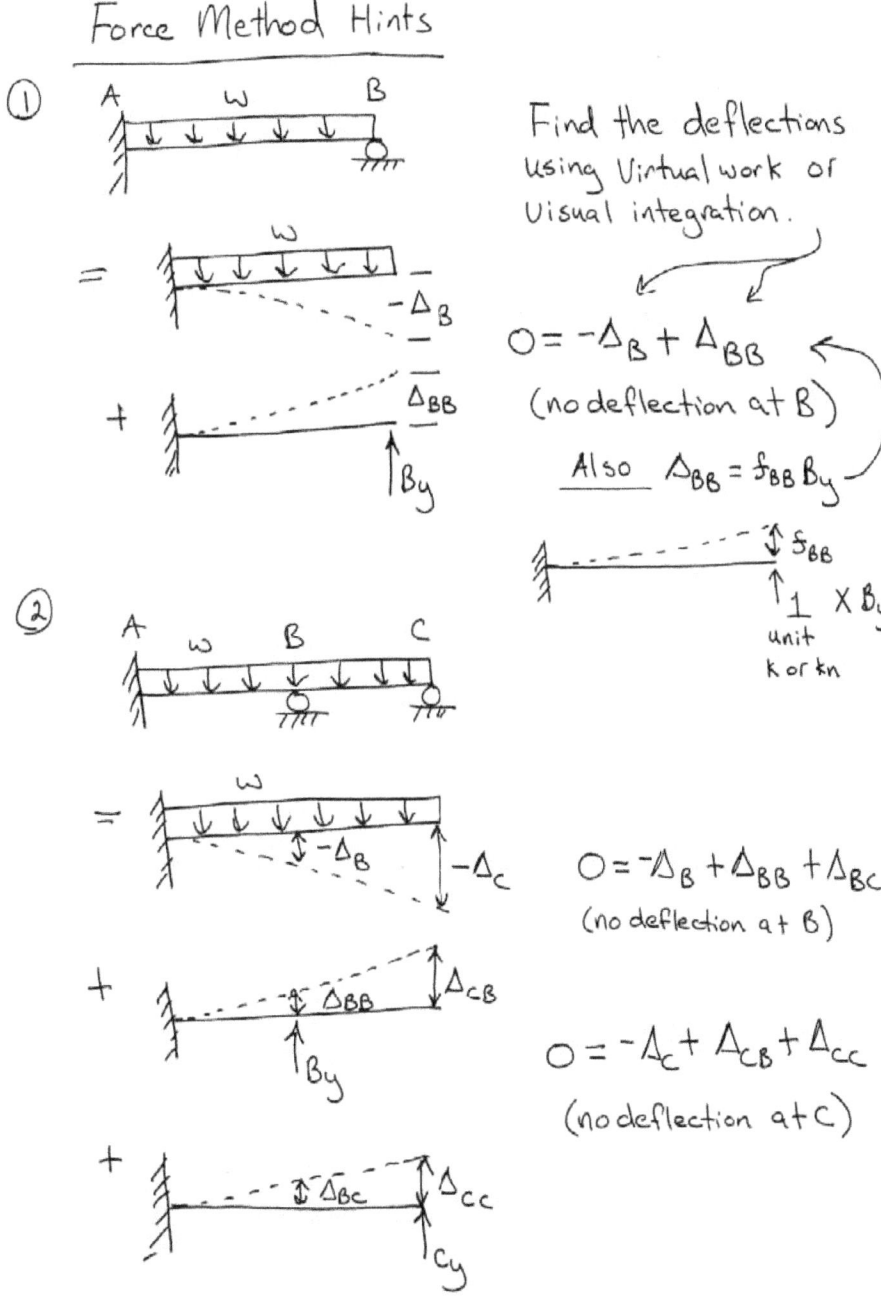

Find the deflections using Virtual work or Visual integration.

$$0 = -\Delta_B + \Delta_{BB}$$

(no deflection at B)

Also $\Delta_{BB} = f_{BB} B_y$

unit k or kn

$$0 = -\Delta_B + \Delta_{BB} + \Delta_{BC}$$

(no deflection at B)

$$0 = -\Delta_C + \Delta_{CB} + \Delta_{CC}$$

(no deflection at C)

Beams

Beam 1

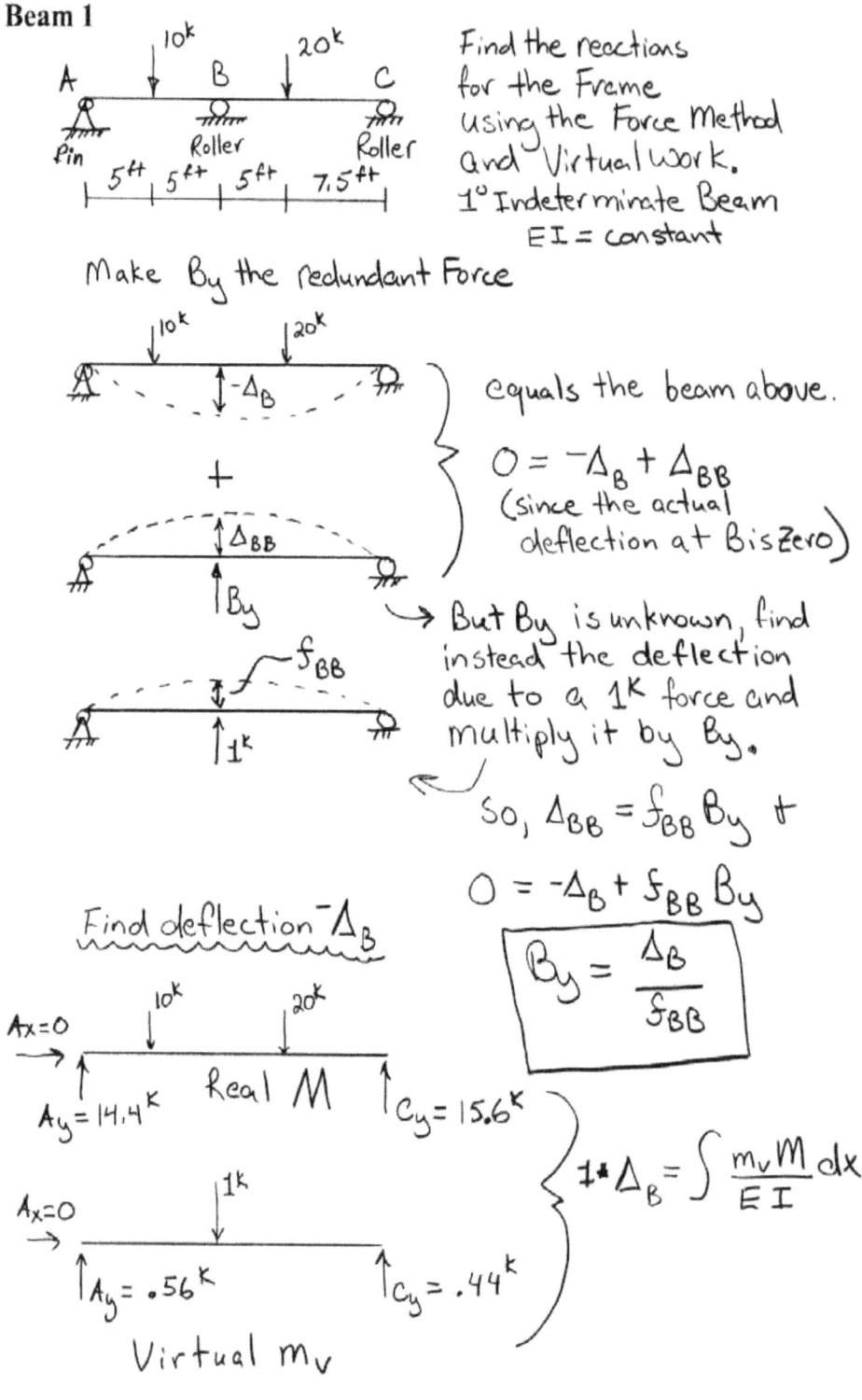

10^k 20^k

A   B   C

Pin   Roller   Roller

5ft   5ft   5ft   7.5ft

Find the reactions for the Frame using the Force Method and Virtual work.
1° Indeterminate Beam
EI = constant

Make B_y the redundant Force

10^k 20^k

$-\Delta_B$

equals the beam above.

+

Δ_{BB}

B_y

$0 = -\Delta_B + \Delta_{BB}$
(since the actual deflection at B is zero)

$-f_{BB}$

1^k

→ But B_y is unknown, find instead the deflection due to a 1^k force and multiply it by B_y.

So, $\Delta_{BB} = f_{BB} B_y$ ↕

$0 = -\Delta_B + f_{BB} B_y$

$$B_y = \frac{\Delta_B}{f_{BB}}$$

Find deflection $-\Delta_B$

10^k 20^k

$A_x = 0$

Real M

$A_y = 14.4^k$ $C_y = 15.6^k$

1^k

$A_x = 0$

$A_y = .56^k$ $C_y = .44^k$

Virtual m_v

$$1^* \Delta_B = \int \frac{m_v M}{EI} dx$$

Beam 1

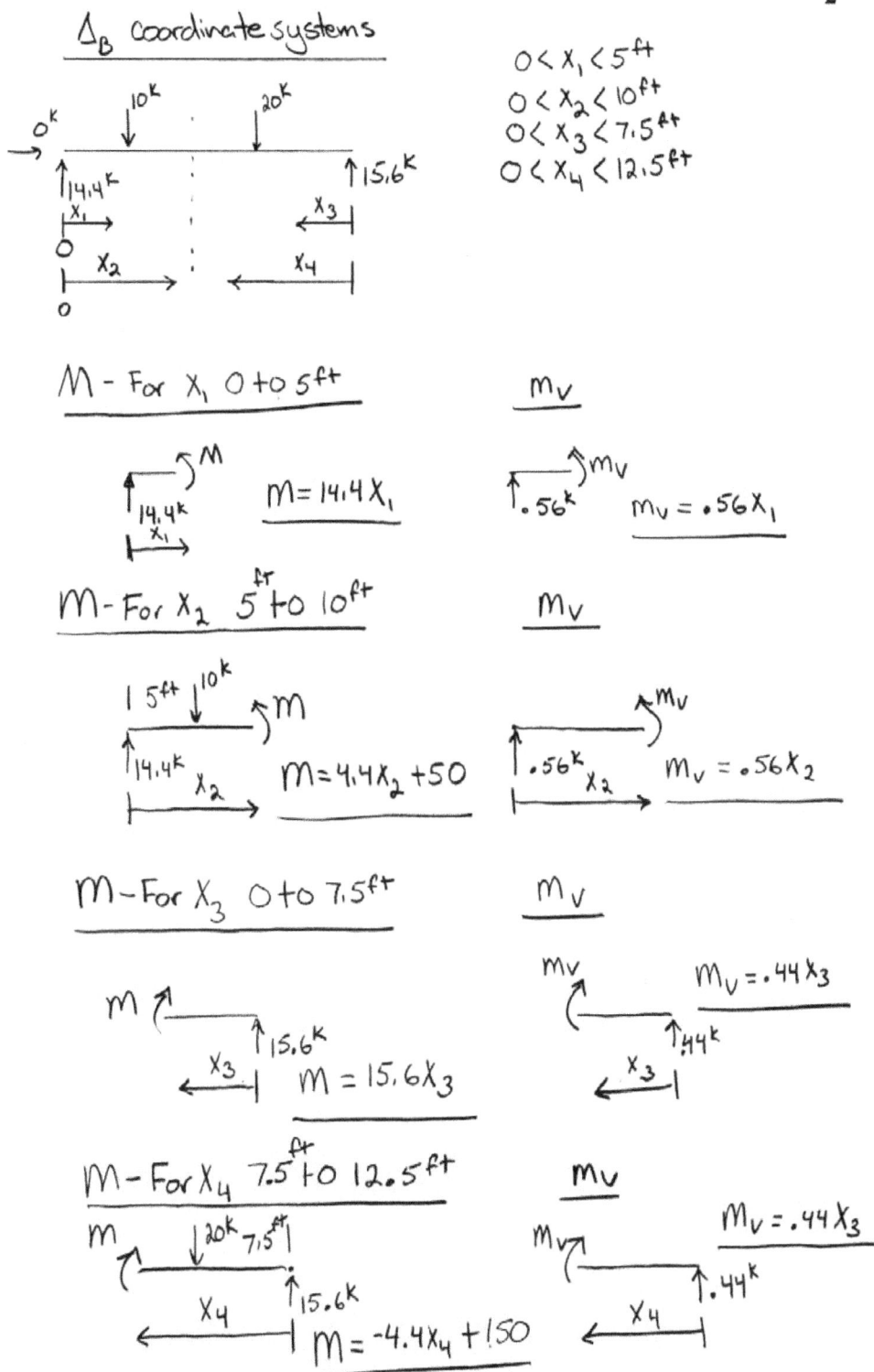

$0 < X_1 < 5 ft$
$0 < X_2 < 10 ft$
$0 < X_3 < 7.5 ft$
$0 < X_4 < 12.5 ft$

M - For X_1 0 to 5ft

m_v

$M = 14.4 X_1$

$m_v = .56 X_1$

M - For X_2 5 ft to 10ft

m_v

$M = 4.4 X_2 + 50$

$m_v = .56 X_2$

M - For X_3 0 to 7.5ft

m_v

$M = 15.6 X_3$

$m_v = .44 X_3$

M - For X_4 7.5 ft to 12.5ft

m_v

$M = -4.4 X_4 + 150$

$m_v = .44 X_3$

$$1 * \Delta_B = \frac{1}{EI}\left[\int_0^5 (.56x_1)(14.4x_1)\,dx + \int_5^{10} (.56x_2)(4.4x_2 + 50)\,dx \right.$$

$$\left. + \int_0^{7.5} (.44x_3)(15.6x_3)\,dx + \int_{7.5}^{12.5} (.44x_4)(-4.4x_4 + 150)\,dx \right]$$

$$\Delta_B = \frac{5372.8}{EI}$$

Find deflection f_{BB}

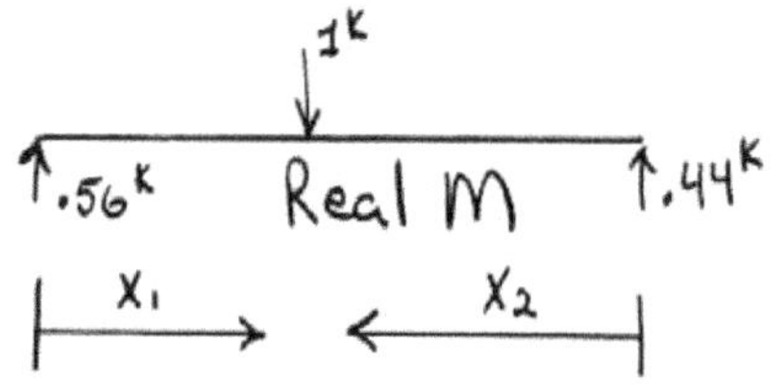

Coordinate systems

$0 < x_1 < 10\text{ft}$
$0 < x_2 < 12.5\text{ft}$

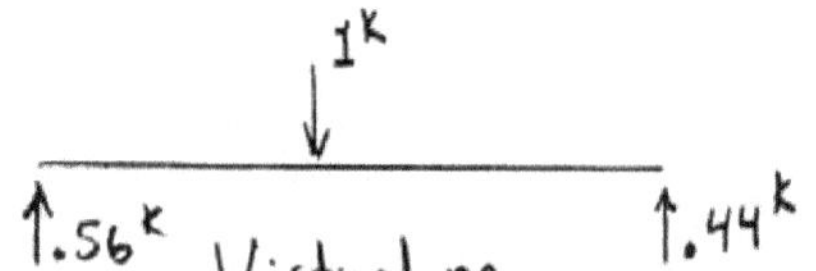

Virtual m_V

M - For x_1 0 to 10^{ft}

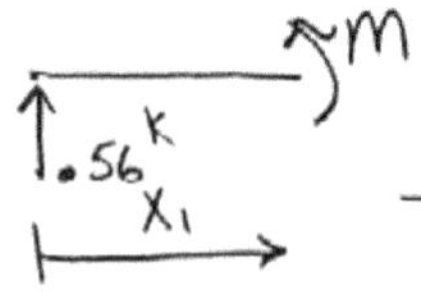

$m = .56x_1$

m_V

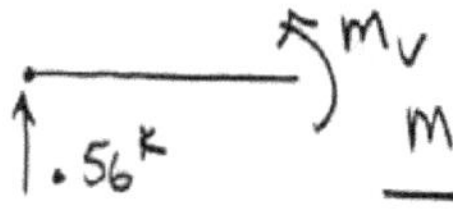

$m_V = .56x_1$

M - For x_2 0 to 12.5^{ft}

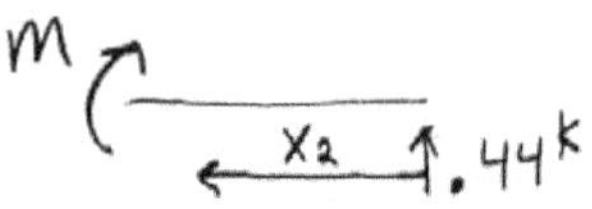

$m = .44x_2$

m_V

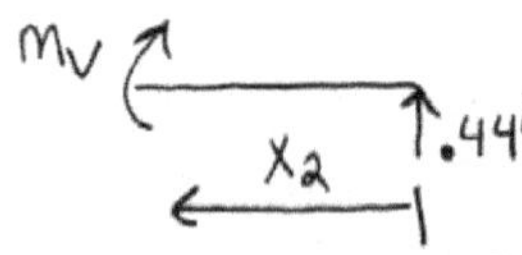

$m_V = .44x_2$

Beam 1

$$1 * f_{BB} = \frac{1}{EI}\left[\int_0^{10}(.56x_1)(.56x_1)\,dx + \int_0^{12.5}(.44x_2)(.44x_2)\,dx\right]$$

$$f_{BB} = \frac{230.575}{EI}$$

Assumed B_y caused a downward deflection, got a positive value for f_{BB} which makes it negative down.

$$0 = -\Delta_B - f_{BB}\,B_y$$

$$B_y = \frac{-\Delta_B}{f_{BB}} = \frac{-\dfrac{5372.8}{EI}}{\dfrac{230.575}{EI}}$$

$A_x = 0$

10^k 20^k

$A_y = 1.5^k$ $B_y = 23.3^k$ $C_y = 5.2^k$ $B_y = 23.3^k \uparrow$

Final Beam after solving for $C_y + A_y$.

Beam 2

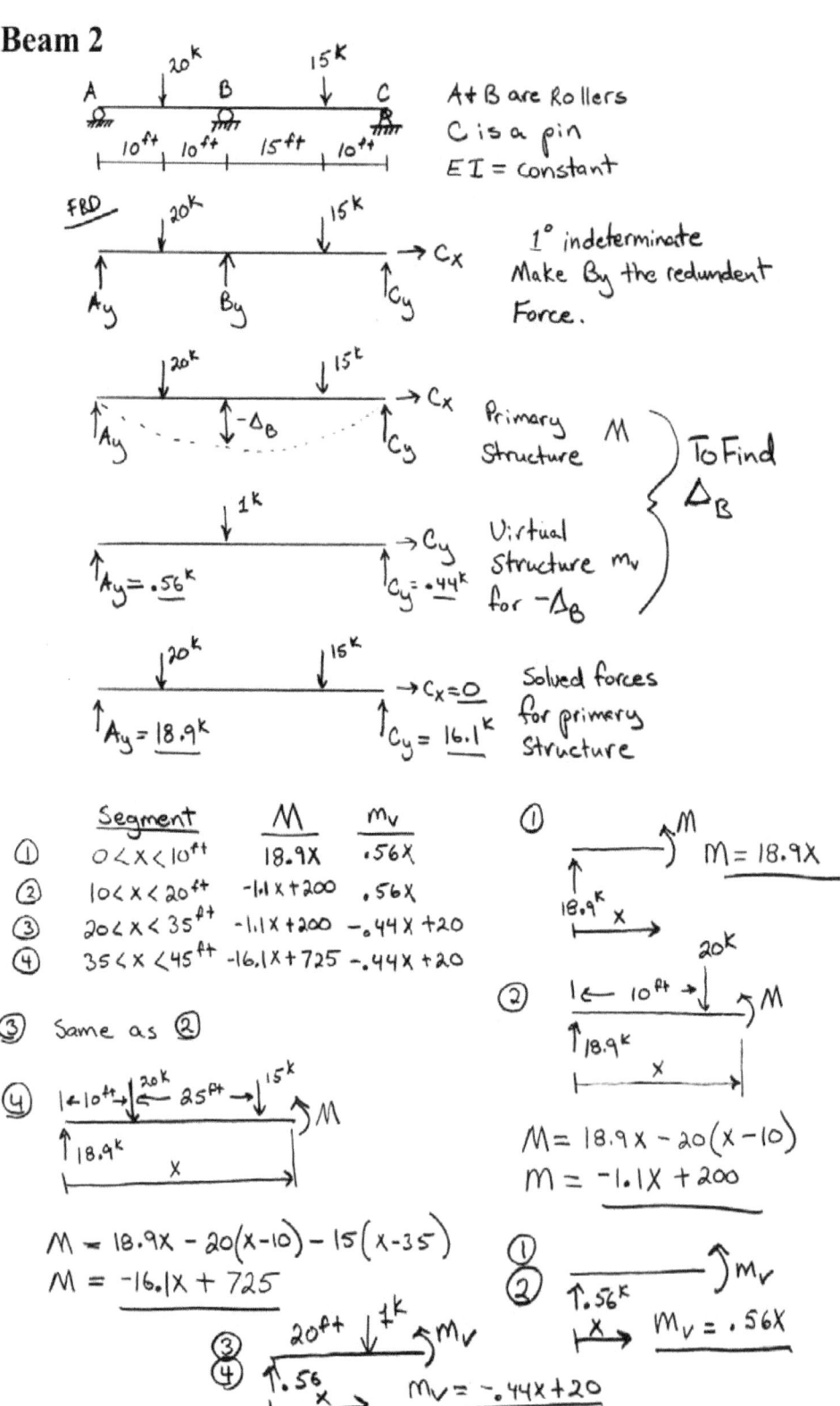

A + B are Rollers
C is a pin
EI = constant

FBD

1° indeterminate
Make B_y the redundent Force.

Primary Structure
Virtual Structure m_v for $-\Delta_B$

$\left.\begin{array}{l} \text{To Find} \\ \Delta_B \end{array}\right\}$

Solved forces for primary Structure

$A_y = 18.9^k$
$C_y = 16.1^k$ $C_x = 0$

Segment		M	m_v
①	$0 < x < 10^{ft}$	$18.9X$	$.56X$
②	$10 < x < 20^{ft}$	$-1.1X + 200$	$.56X$
③	$20 < x < 35^{ft}$	$-1.1X + 200$	$-.44X + 20$
④	$35 < x < 45^{ft}$	$-16.1X + 725$	$-.44X + 20$

③ Same as ②

④ $M = 18.9X - 20(x-10) - 15(x-35)$
   $M = -16.1X + 725$

① $M = 18.9X$

② $M = 18.9X - 20(X-10)$
   $M = -1.1X + 200$

$\begin{array}{c} ① \\ ② \end{array}$ $m_v = .56X$

$\begin{array}{c} ③ \\ ④ \end{array}$ $m_v = -.44X + 20$

Beam 2

$$1^k \cdot \Delta_B = \int \frac{M \, m_v}{EI} \, dx$$

$$1^k \cdot \Delta_B = \int_0^{10} \frac{(18.9x)(.56x)}{EI} \, dx + \int_{10}^{20} \frac{(-1.1x+200)(.56x)}{EI} \, dx$$

$$+ \int_{20}^{35} \frac{(-1.1x+200)(-.44x+20)}{EI} \, dx + \int_{35}^{45} \frac{(-16.1x+725)(-.44x+20)}{EI} \, dx$$

$$\Delta_B = \frac{41,676.5}{EI}$$

$$0 = -\Delta_B + \Delta_{BB}$$

Primary

$$\Delta_{BB} = -f_{BB} B_y$$

M values

Virtual

m_v values

$$1 \ast f_{BB} = \int_0^{20} \frac{(.56x)(.56x)}{EI} \, dx + \int_{20}^{45} \frac{(-.44x+20)(-.44x+20)}{EI} \, dk$$

$$f_{BB} = \frac{1900.6}{EI}$$

$$B_y = \frac{\Delta_{BB}}{f_{BB}} = \frac{\dfrac{41,676.5}{EI}}{\dfrac{1900.6}{EI}} = 21.93^k \uparrow$$

$$C_x = 0^k$$

$$C_y = 6.36^k \uparrow$$

$$A_y = 6.71^k \uparrow$$

Beam 3

Force Method — Moment as redundant (Using Book Formulas)

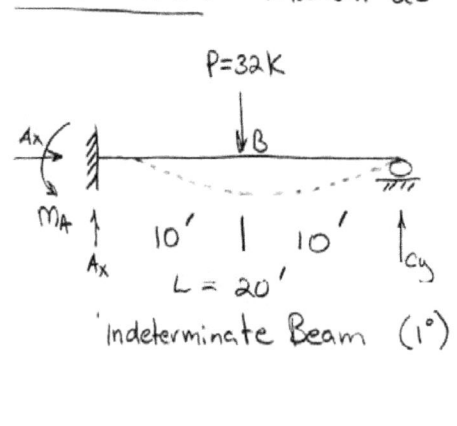

$P = 32K$

A_x

$\downarrow B$

$M_A \uparrow$ 10' | 10' $\uparrow_{c_y}$

A_x

$L = 20'$

Indeterminate Beam (1°)

$E = 30,000\ ksi$

$I = 512\ in^4$

$$\boxed{\Theta_A + f_{AA}\, M_A = 0}$$

Book formula

$P = 32^k$

$A_x = 0$

$\rightarrow$

$\uparrow_{\Theta_A}$

$A_y = 16^k$ $C_y = 6^k$

Primary Beam Subjected
to External Loading

$$\Theta_A = -\frac{PL^2}{16EI} = \frac{-32(20)^2}{16\left(30,000\right)\left(\frac{512}{144}\right)}$$

$$\boxed{\Theta_A = -0.0075\ rad}$$

$$f_{AA} = \frac{(m)\,L}{3EI} = \frac{1(20)}{3\left(30000\right)\left(\frac{512}{144}\right)}$$

$$\boxed{f_{AA} = .0000625\ \frac{rad}{k\,ft}}$$

+

$A_x = 0$ $1\,kft$

$\rightarrow$ $\downarrow \xi_M$ C $\times m_A$

$\uparrow A_y = \frac{1}{20}^k$ $C_y = \frac{1}{20}^k$

Primary Beam Loaded
with redundant M_A

$$M_A = \frac{0.0075}{0.000625}$$

$$\boxed{M_A = 120\,k\text{-}ft\ \circlearrowright}$$

$$\boxed{\begin{array}{l} C_y = 10^k \\[4pt] A_y = 22^k \\[4pt] A_x = 0 \end{array}}$$

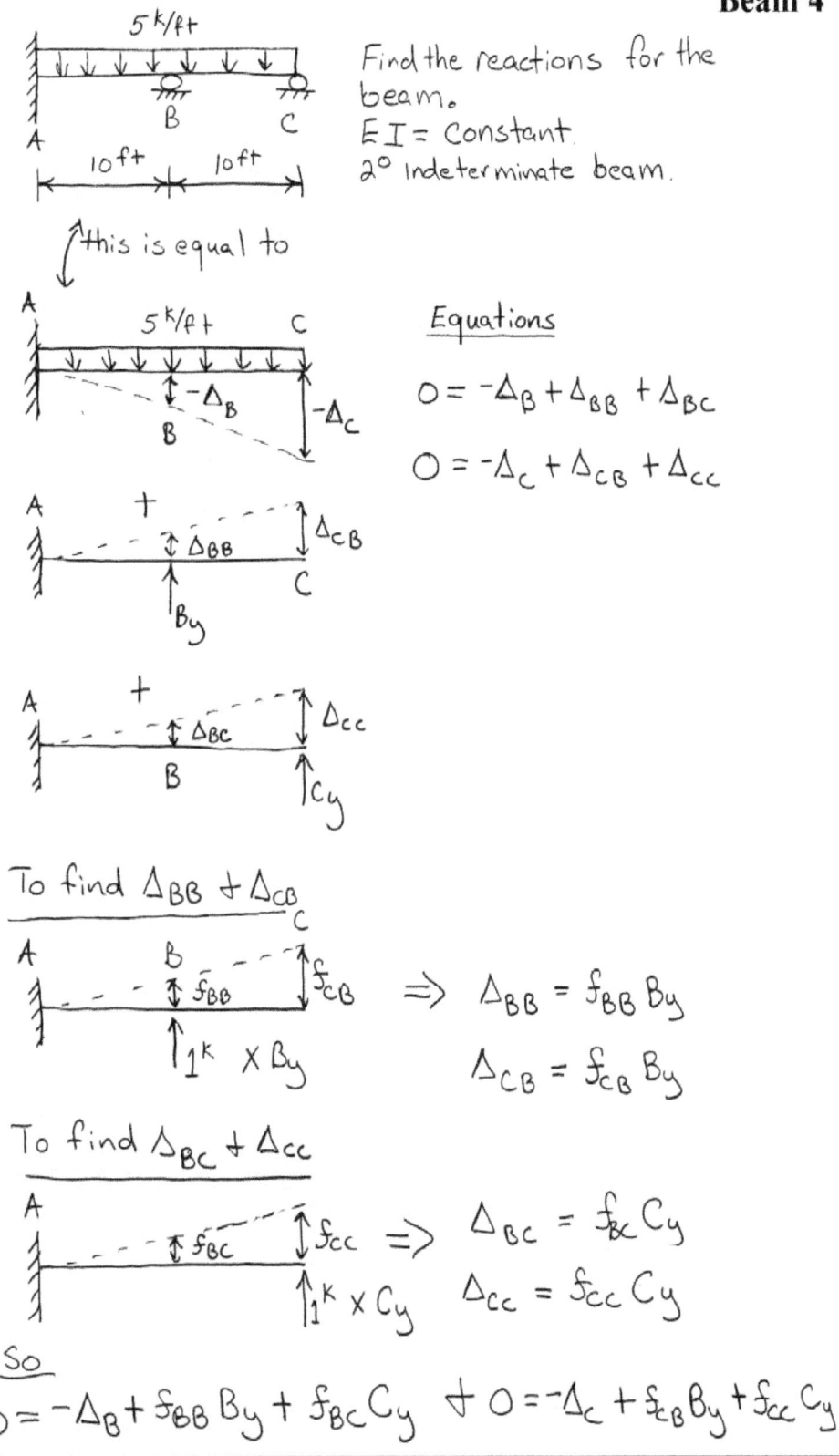

Find the reactions for the beam.
EI = Constant.
$2°$ Indeterminate beam.

this is equal to

Equations

$$0 = -\Delta_B + \Delta_{BB} + \Delta_{BC}$$

$$0 = -\Delta_C + \Delta_{CB} + \Delta_{CC}$$

To find Δ_{BB} & Δ_{CB}

$$\Rightarrow \Delta_{BB} = f_{BB} B_y$$

$$\Delta_{CB} = f_{CB} B_y$$

To find Δ_{BC} & Δ_{CC}

$$\Rightarrow \Delta_{BC} = f_{BC} C_y$$

$$\Delta_{CC} = f_{CC} C_y$$

So

$$0 = -\Delta_B + f_{BB} B_y + f_{BC} C_y \qquad \& \qquad 0 = -\Delta_C + f_{CB} B_y + f_{CC} C_y$$

Beam 4

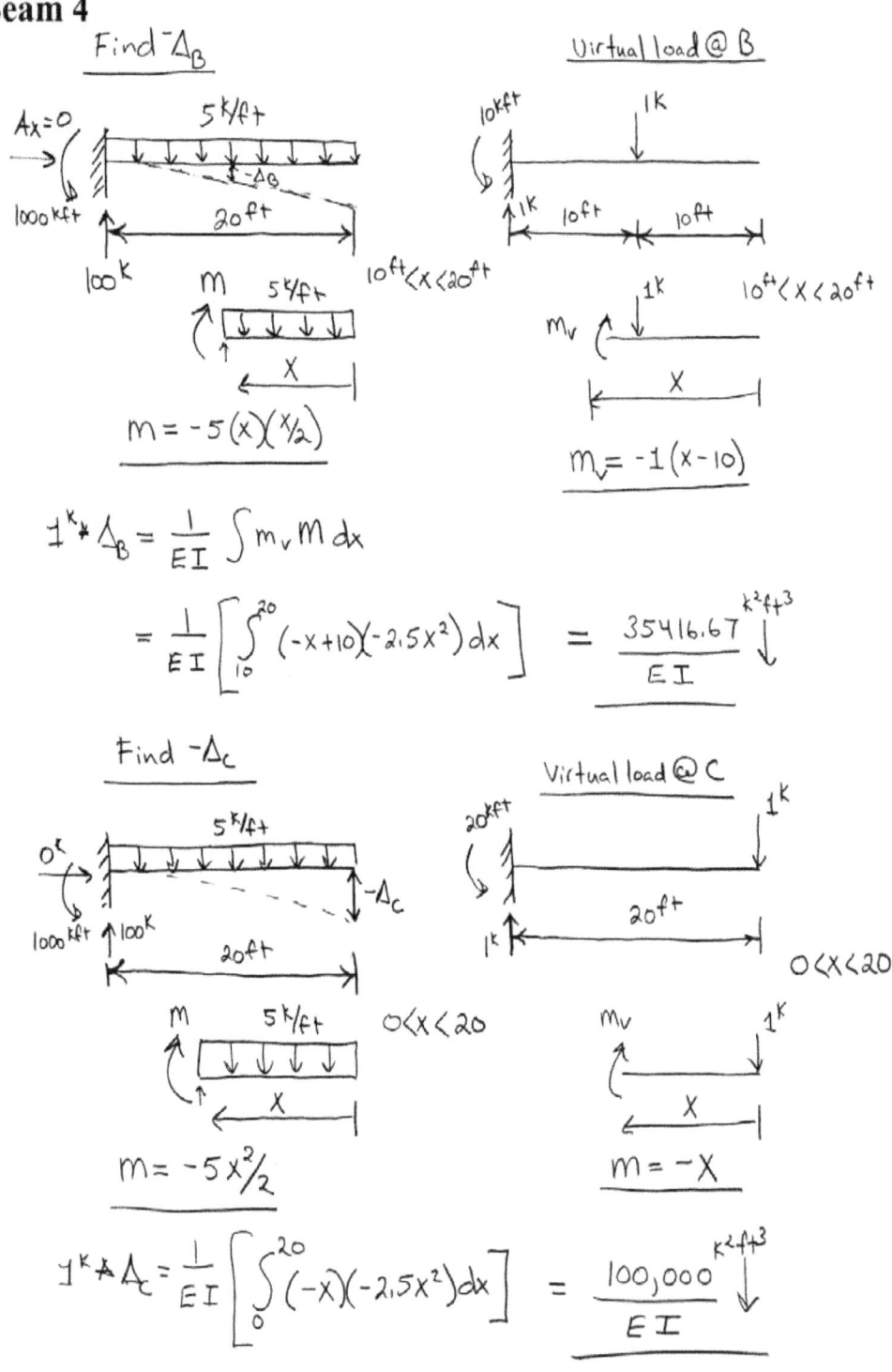

Find $-\Delta_B$

$A_x = 0$

$5^k/ft$

$-\Delta_B$

1000^{kft}

$20 ft$

100^k

m $5^k/ft$ $10ft < x < 20ft$

x

$$m = -5(x)(x/2)$$

$$1^k * \Delta_B = \frac{1}{EI} \int m_v m \, dx$$

$$= \frac{1}{EI}\left[\int_{10}^{20}(-x+10)(-2.5x^2)\,dx\right] = \frac{35416.67}{EI} \downarrow \; ^{k^2ft^3}$$

Virtual load @ B

10^{kft} $1K$

$1K$ $10ft$ $10ft$

m_v $1K$ $10ft < x < 20ft$

x

$$m_v = -1(x-10)$$

Find $-\Delta_C$

0^c

$5^k/ft$

$-\Delta_C$

1000^{kft} 100^k

$20 ft$

m $5^k/ft$ $0 < x < 20$

x

$$m = -5x^2/2$$

$$1^k * \Delta_C = \frac{1}{EI}\left[\int_{0}^{20}(-x)(-2.5x^2)\,dx\right] = \frac{100,000}{EI} \downarrow \; ^{k^2ft^3}$$

Virtual load @ C

20^{kft} 1^k

$20ft$

1^k $0 < x < 20$

m_v 1^k

x

$$m = -x$$

Beam 4

Find f_{BB}

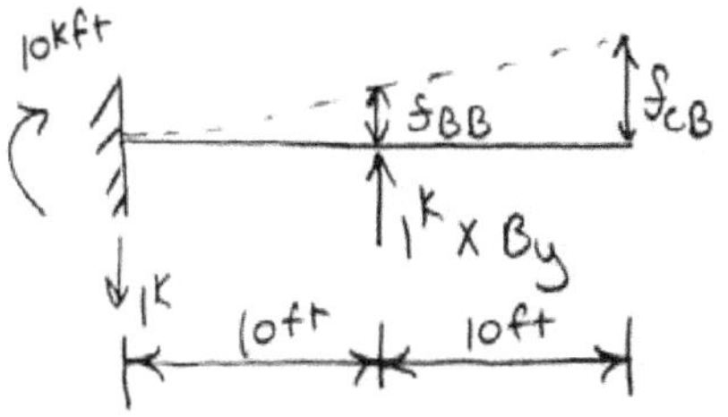

10kft

f_{BB} f_{CB}

$1^k \times B_y$

1k

10ft   10ft

Virtual load @ B

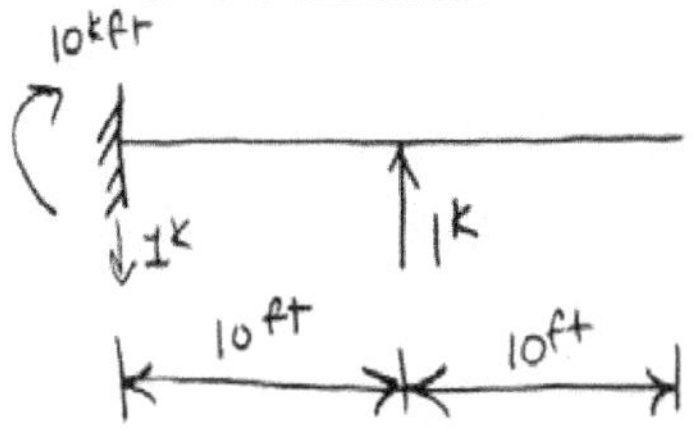

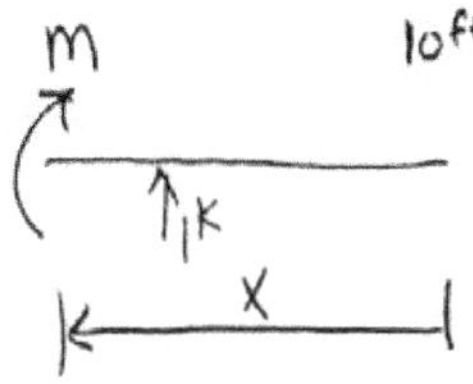

10kft

1k   1k

10ft   10ft

m $10\text{ft} < x < 20\text{ft}$

1k

x

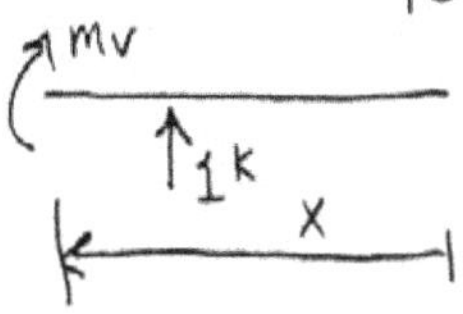

$m = x - 10$

m_v $10\text{ft} < x < 20\text{ft}$

$1k$

x

$m_v = x - 10$

$$1^k * f_{BB} = \frac{1}{EI}\left[\int_{10}^{20}(x-10)^2 dx\right] = \frac{333.33}{EI} \; \overset{k^2 ft^3/k}{\uparrow}$$

Find f_{CB}

Real same as above

$m = x - 10$

20kft Virtual load @ C

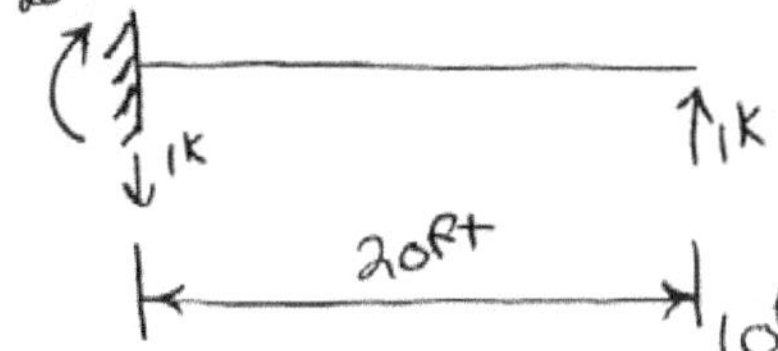

1k   1k

20ft

$10\text{ft} < x < 20\text{ft}$

m_v

1k

x

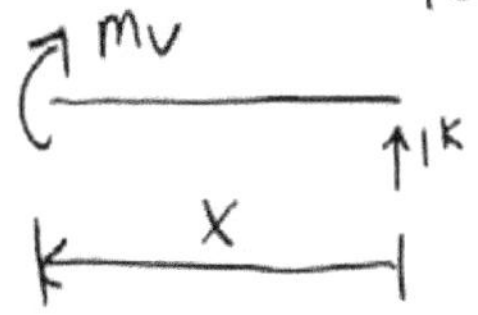

$m_v = x$

$$1^k * f_{CB} = \frac{1}{EI}\left[\int_{10}^{20}(x-10)(x)dx\right]$$

$$= \frac{833.33}{EI} \; \overset{k^2 ft^3/k}{\uparrow}$$

Find f_{cc}

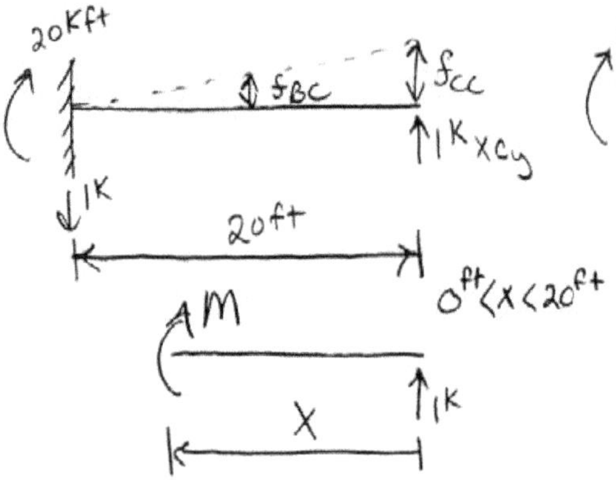

20kft

f_{BC} f_{cc}

$1^k x_{c_y}$

1K

20ft

$0^{ft} < x < 20^{ft}$

M

X 1K

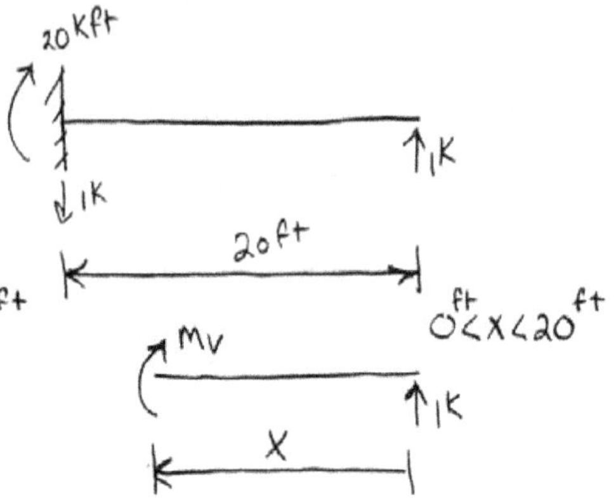

$m = X$

$$1^k * f_{cc} = \frac{1}{EI}\left[\int_0^{20} (x)(x)dx\right] = \frac{2666.67}{EI}^{k^2 ft^3/k} \uparrow$$

Virtual load @ C

20kft

1K

1K

20ft

$0 < x < 20^{ft}$

m_v

X 1K

$m_v = X$

Find f_{BC}

Real system same as above

$m = X$ but $10^{ft} < x < 20^{ft}$

$$1^k * f_{BC} = \frac{1}{EI}\left[\int_{10}^{20} (x-10)(x)dx\right]$$

$$= \frac{833.33}{EI}^{k^2 ft^3/k} \uparrow$$

Virtual load @ B

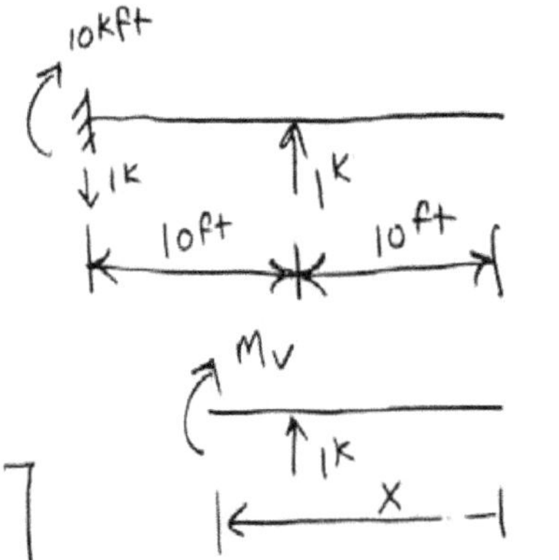

10kft

1K 1K

10ft 10ft

m_v

1K

X

$m_v = X - 10$

Beam 4

two equations, two unknowns.

$$0 = -\Delta_B + f_{BB} B_y + f_{BC} C_y$$

$$0 = -\Delta_C + f_{CB} B_y + f_{CC} C_y$$

$\underline{so}$

$$0 = -\frac{3541667}{EI} + \frac{333.33}{EI} B_y + \frac{833.33}{EI} C_y$$

$$0 = \frac{-100,000}{EI} + \frac{833.33}{EI} B_y + \frac{2666.67}{EI} C_y$$

$$B_y = \underline{\underline{57.1}}^k$$

$$C_y = \underline{\underline{19.6}}^k$$

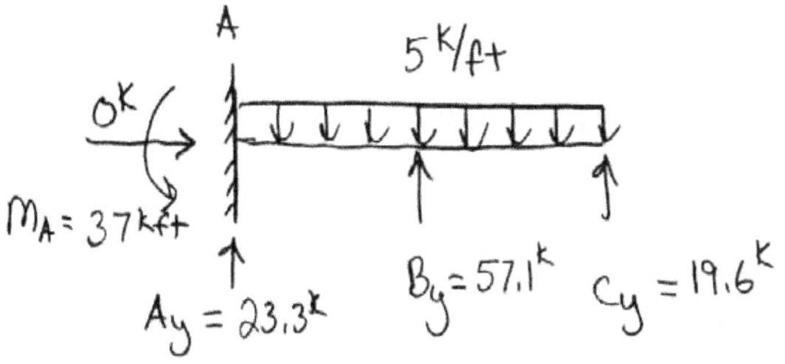

A

$5^{k/ft}$

0^k

$M_A = 37 k ft$

$A_y = 23.3^k$ $B_y = 57.1^k$ $C_y = 19.6^k$

Frames

Frame 1

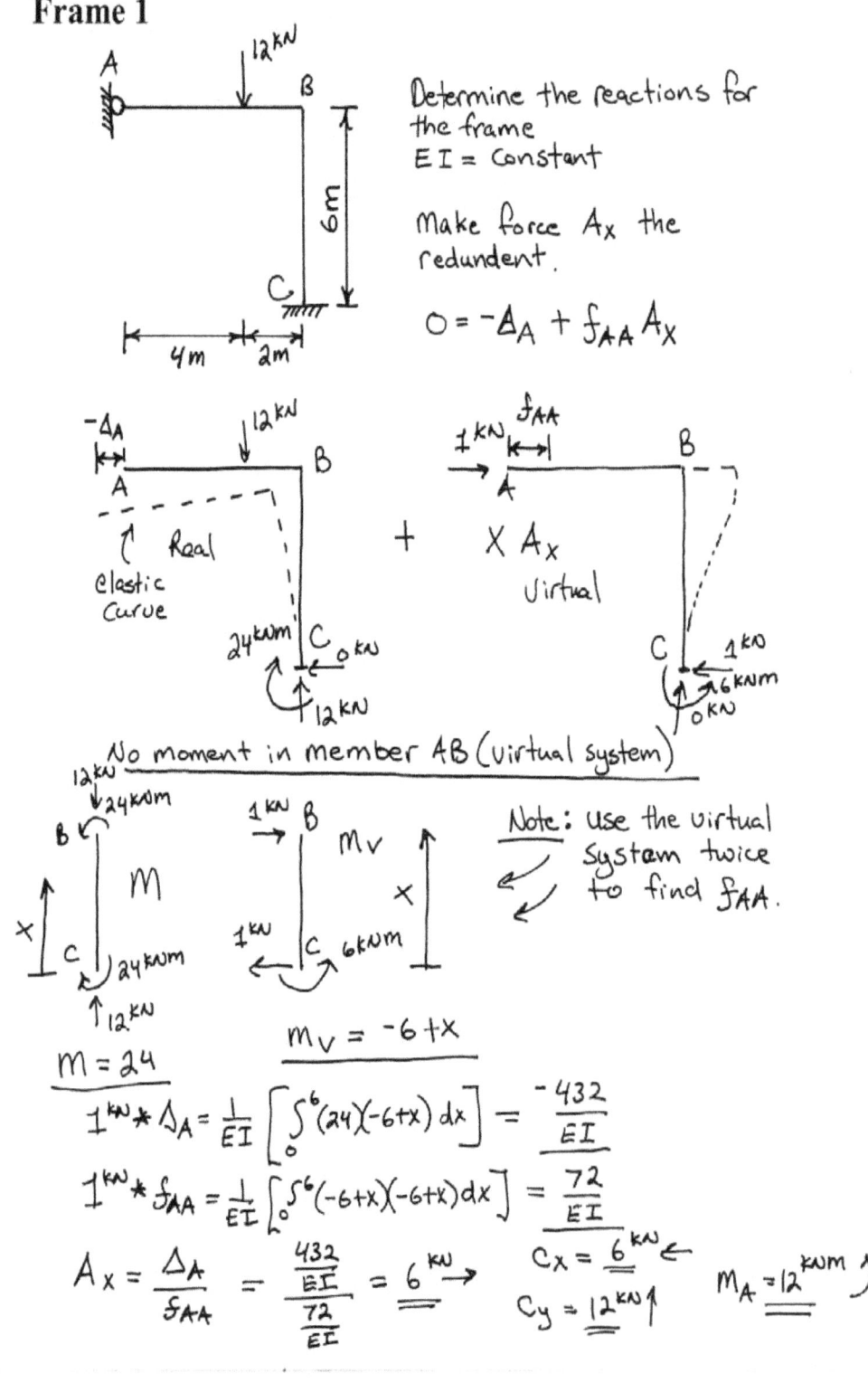

Determine the reactions for
the frame
$EI = $ Constant

Make force A_x the
redundent.

$$0 = -\Delta_A + f_{AA} A_x$$

No moment in member AB (virtual system)

Note: use the virtual
System twice
to find f_{AA}.

$m = 24$

$\underline{m_v = -6+x}$

$$1^{kN} * \Delta_A = \frac{1}{EI} \left[\int_0^6 (24)(-6+x)\, dx \right] = \frac{-432}{EI}$$

$$1^{kN} * f_{AA} = \frac{1}{EI} \left[\int_0^6 (-6+x)(-6+x)\, dx \right] = \frac{72}{EI}$$

$$A_x = \frac{\Delta_A}{f_{AA}} = \frac{\frac{432}{EI}}{\frac{72}{EI}} = \underline{6 \xrightarrow{kN}} \qquad \underline{C_x = 6^{kN} \leftarrow} \qquad \underline{M_A = 12^{kNm} \curvearrowright}$$

$$\underline{C_y = 12^{kN} \uparrow}$$

Force Method using Visual Integration

Beam 1

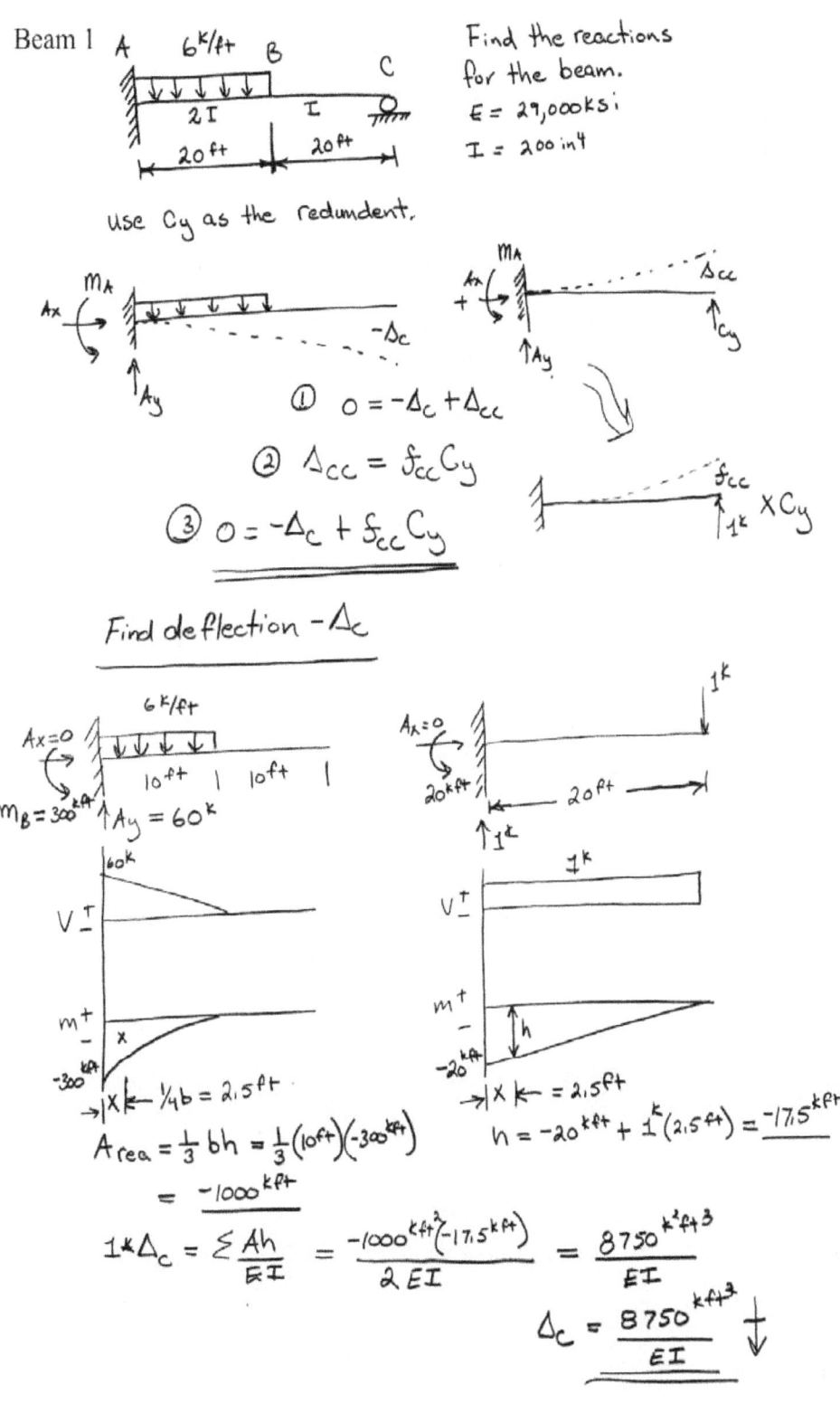

Find the reactions for the beam.

$E = 29,000 \text{ ksi}$

$I = 200 \text{ in}^4$

use C_y as the redundent.

① $0 = -\Delta_c + \Delta_{cc}$

② $\Delta_{cc} = f_{cc} C_y$

③ $0 = -\Delta_c + f_{cc} C_y$

Find deflection $-\Delta_c$

$A_x = 0$

$m_B = 300^{kA}$ $A_y = 60^k$

$\rightarrow |x \leftarrow \frac{1}{4}b = 2.5 ft$

$A_{rea} = \frac{1}{3} bh = \frac{1}{3}(10 ft)(-300^{kA})$

$= -1000^{kft}$

$A_x = 0$

$\rightarrow |x \leftarrow = 2.5 ft$

$h = -20^{kft} + 1^k (2.5 ft) = -17.5^{kft}$

$1 \ast \Delta_c = \sum \frac{Ah}{EI} = \frac{-1000^{kft^2}(-17.5^{kft})}{2 EI} = \frac{8750^{k^2 ft^3}}{EI}$

$\Delta_c = \frac{8750^{kft^3}}{EI} \downarrow$

Beam 1

Find deflection f_{cc}

Left diagram: cantilever beam with 0^k, 20^{kft} moment, 1^k down, 1^k up, $10ft$ and $10ft$ spans.

V diagram, m diagram with -1^k, 20^{kft}, A_1, 10^{kft}, A_3, A_2 $10ft$ | $10ft$.

centriod

3 Areas A_1 @ $10/3'$

$x \rightarrow$ A_2 @ $5'$

A_3 @ $10' + 10/3'$

Right diagram: 0^k, 20^{kft}, 1^k down, 1^k up, $10ft$ and $10ft$.

V diagram, m diagram with -1^k, 20^{kft}, h_2, 10^{kft}, h_3, h_1, $10ft$ | $10ft$.

$$1^k * f_{cc} = \sum \frac{Ah}{EI} = \frac{A_1 h_1}{2EI} + \frac{A_2 h_2}{2EI} + \frac{A_3 h_3}{EI}$$

$$= \frac{\frac{1}{2}(10^{kft})(10ft)(16.67)^{kft}}{2EI} + \frac{(10^{kft})(10ft)(15^{kft})}{2EI} + \frac{\frac{1}{2}(10^{kft})(10ft)(6.67)^{kft}}{EI}$$

$$= \frac{416.75}{EI} + \frac{750}{EI} + \frac{333.5}{EI} = \frac{1500.25^{k^2 ft^3}}{EI}$$

$$f_{cc} = \frac{1500.25^k ft^3}{EI}$$

$$0 = -\Delta_c + f_{cc} C_y \Rightarrow 0 = \frac{-8750^{kft^3}}{EI} + \frac{1500.25^{kft^3}}{EI} C_y$$

$$C_y = 5.83^k \qquad A_x = 0^k$$

$$A_y = 54.17^k \qquad M_A = 183^{kft}$$

Beam 2

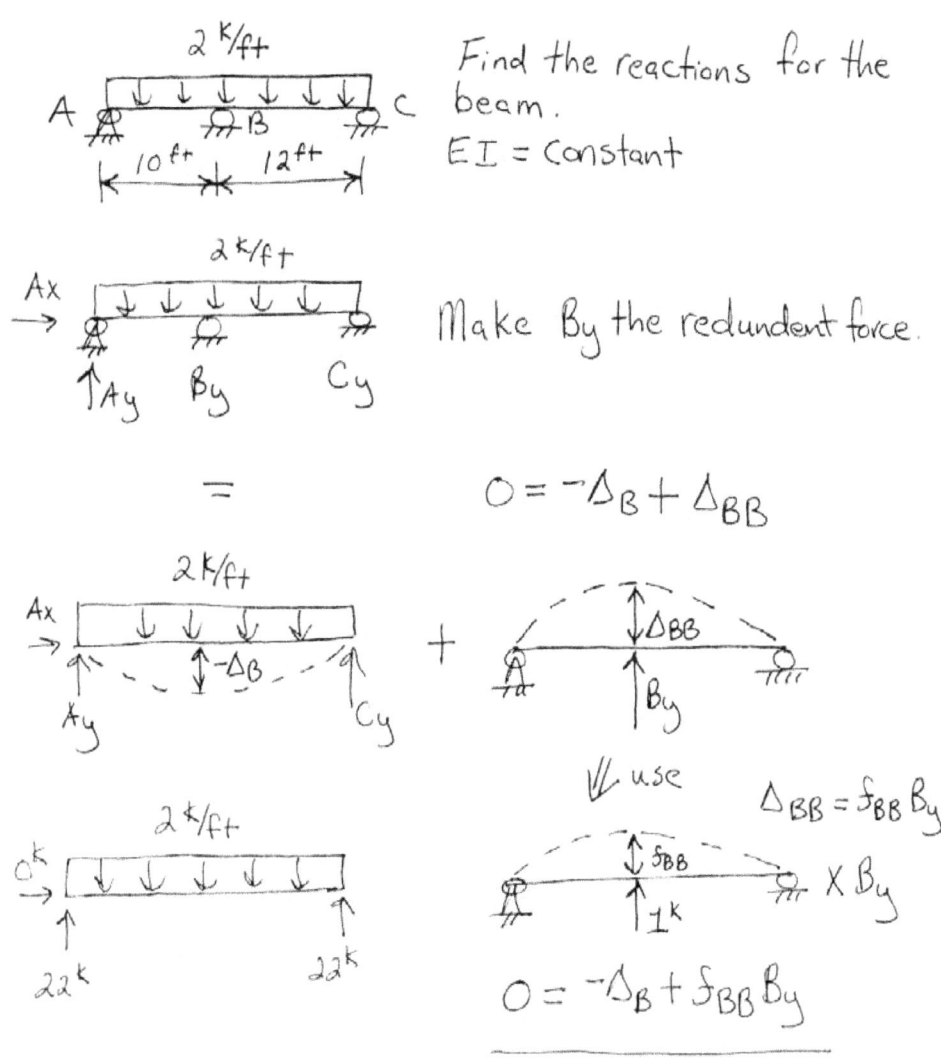

Find the reactions for the beam.

$EI = $ Constant

Make B_y the redundent force.

$0 = -\Delta_B + \Delta_{BB}$

$\text{\Large +}$

$\Downarrow$ use

$\Delta_{BB} = f_{BB} B_y$

$\times B_y$

$0 = -\Delta_B + f_{BB} B_y$

o^k

$2^k/ft$

22^k 22^k

$22ft$

22^k

V_-^+

$121 kft$ -22^k

m_-^+ $11 ft$ $11 ft$

1^k

$.55 K$ $.45^k$

$10ft$ $12ft$

$.55$ k

V_-^+ $-.45^k$

$5.5 kft$

m_-^+ $10ft$ $12ft$

↑ hard to work with
So use ⟩ Cantilever parts

Assume B fixed

22^k 22^k

$10 ft$ $12 ft$
$22 k$

V_-^+ -22^k

A_1 $220 kft$ $264 kft$ A_2

m_-^+ $10ft$ $12ft$

$$1^k * \Delta_B = \frac{\Sigma Ah}{EI}$$

$+$

Assume B fixed

$2^k/ft$ $2^k/ft$

24^k

V_-^+
-20^k

m_-^+ A_3 -100 A_4

$-144 kft$

$10ft$ $12ft$

$$1^k * \Delta_B = \frac{1}{EI}\left[\frac{1}{2}(10ft)(220^{kft})(3.64\ kft) + \frac{1}{2}(12ft)(264^{kft})(3.64^{kft}) \right.$$

$$\left. + \frac{1}{3}(+10^{ft})(-100^{kft})(4.09^{kft}) + \frac{1}{3}(12^{ft})(-144^{kft})(4.09\ kft) \right]$$

(A_1, h_1, A_2, h_2, A_3, h_3, A_4, h_4 labels)

$$\Delta_B = \frac{6050\ kft^3}{EI}$$

Find Δ_{BB}

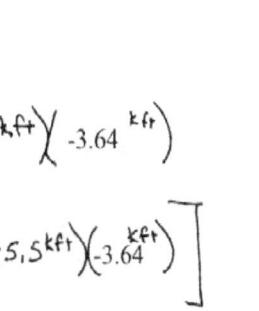

$$1 * f_{BB} = \sum \frac{Ah}{EI}$$

$$= \frac{1}{EI}\left[\frac{1}{2}(10^{ft})(-5.5^{kft})(-3.64^{kft})\right.$$

$$\left. + \frac{1}{2}(12^{ft})(-5.5^{kft})(-3.64^{kft})\right]$$

$$f_{BB} = \frac{220 \quad kft^3/k}{EI}$$

$f_{BB} \downarrow \ B_1$

$1^k \ X \ B_y$

$.55^k$ .... $.45^k$

1^k

$\leftarrow 10^{ft} \rightarrow \leftarrow 12^{ft} \rightarrow$

$.45^k$

V^+_-

$-.55^k$

m^+_- A_1 A_2

-5.5^{kft}

$$0 = -\Delta_B + f_{BB} B_y$$

$.55^k$ .... $.45^k$

1^k

$$B_y = \frac{\Delta_B}{f_{BB}}$$

$.45^k$

V^+_-

M^+_- h_1 h_2

-5.5^{kft}

$$B_y = \frac{\dfrac{6050}{EI}}{\dfrac{220}{EI}} = 27.5^k$$

$2^k/ft$

$0^k \rightarrow$ ↓↓↓↓↓↓

7^k 27.5^k 9.5^k

$\leftarrow 10^{ft} \rightarrow \leftarrow 12^{ft} \rightarrow$

Final Reactions

Frame 1

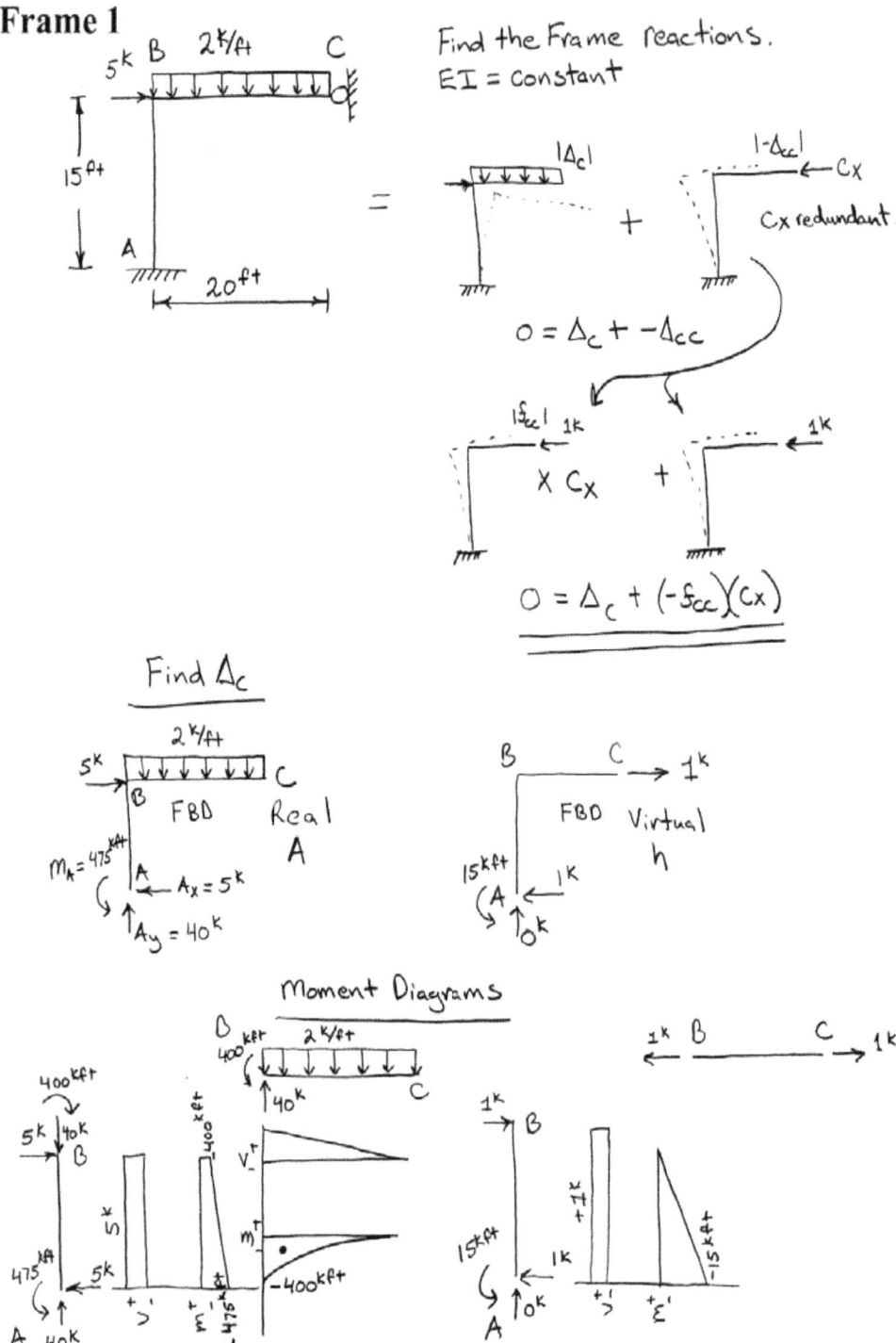

Find the Frame reactions.
$EI = $ constant

$0 = \Delta_c + -\Delta_{cc}$

$0 = \Delta_c + (-f_{cc})(C_x)$

Find Δ_c

FBD Real

FBD Virtual

Moment Diagrams

Frame 1

$\Delta_c = \sum \dfrac{Ah}{EI}$ break AB into 2 Areas

$$\Delta_c = \dfrac{\overset{A_1}{15^{ft}}\overset{}{\left(-400^{kft}\right)}\overset{h_1}{\left(-7.5^{kft}\right)}}{EI} + \dfrac{\frac{1}{2}\overset{A_2}{\left(15^{ft}\right)}\left(-75^{kft}\right)\overset{h_2}{\left(-10^{kft}\right)}}{EI}$$

$$\Delta_c = \dfrac{50,625 \;\; k^2 ft^3}{EI}$$

Find f_{cc}

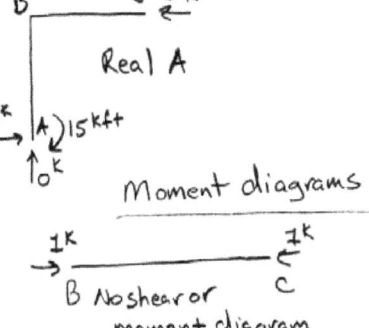

Real A

Moment diagrams

B No shear or moment diagram

virtual h

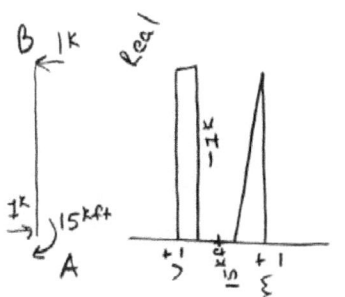

Real

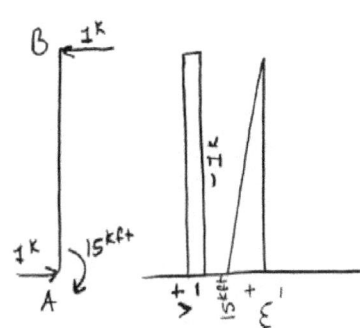

Frame 1

$$f_{cc} = \sum \dfrac{Ah}{EI} = \dfrac{\frac{1}{2}(15^{ft})(15^{kft})(10^{kft})}{EI} = \dfrac{1125\,{}^{k^2ft^3}/k}{EI}$$

$$0 = \Delta_c - f_{cc}\, C_x$$

$$C_x = \dfrac{\Delta_c}{f_{cc}} = \dfrac{\dfrac{50,625\,{}^{k^2ft^3}}{EI}}{\dfrac{1125\,k^2ft^3/k}{EI}} = \underline{\underline{45^k}} \leftarrow$$

$$C_x = \underline{\underline{45^k}} \leftarrow$$

$$A_x = \underline{\underline{40^k}} \rightarrow$$

$$A_y = \underline{\underline{40^k}} \uparrow$$

$$M_A = \underline{\underline{200^{kft}}} \circlearrowright$$

A

12^{KN} B

4m 12m

6m

C

Find the reactions
E I = constant

Frame 2

$$0 = -\Delta_A + f_{AA} A_x$$

A values

12^{KN}

$|-\Delta_A|$ B

C $\leftarrow$ kN

$\curvearrowleft$ 24 kNm

12 kN

+

h values

1^{KN} A
$\rightarrow$

$|f_{AA}|$ B

$f_{AA} A_x$

C $\leftarrow$ 1kN

$\curvearrowleft$ 6 kNm

0 kN

> No shear or moment
> diagram for AB.

12^{KN}

24 kNm

24 kNm

12 kN

24 kNm moment

A_1

•

$+ \Sigma$

$+ \Sigma$

1 kN
$\rightarrow$

1 kN
6 kNm

1 kN

A_2

-6 kNm

$+ \Sigma$

$+ \Sigma$

$$\Delta_A = \Sigma \frac{Ah}{EI} = \frac{24^{kNm}(6m)(3^{kNm})}{EI} = \frac{-432}{EI} \, kN^2m^3$$

$$f_{AA} = \Sigma \frac{Ah}{EI} = \frac{\frac{1}{2}(-6^{kNm})(6m)(-4^{kN}m)}{EI} = \frac{72}{EI} \, kN^2m^3$$

$$A_x = \frac{\Delta_A}{f_{AA}} = \frac{\frac{432}{EI}}{\frac{72}{EI}} = \underline{\underline{6}}^{kN} \rightarrow$$

Frame 2

6 kN →   ↓12 kN

⤹ Cx

↑ ↺ m

↑ Cy

$\xrightarrow{+} \ \Sigma F_x = 0$

$-C_x + 6^{kN} = 0$

$C_x = \underline{\underline{6^{kN}}}$

$\uparrow^+ \ \Sigma F_y = 0$

$-12^{kN} + C_y = 0$

$C_y = \underline{\underline{12^{kN}}}$

$\circlearrowleft^{(+)} \ \Sigma M_C = 0$

$12^{\frac{kN}{}}(2m) - 6^{kN}(6m) - m = 0$

$m = \underline{-12^{kNm}}$

$m = \underline{\underline{12^{kNm}}} \ \circlearrowright$

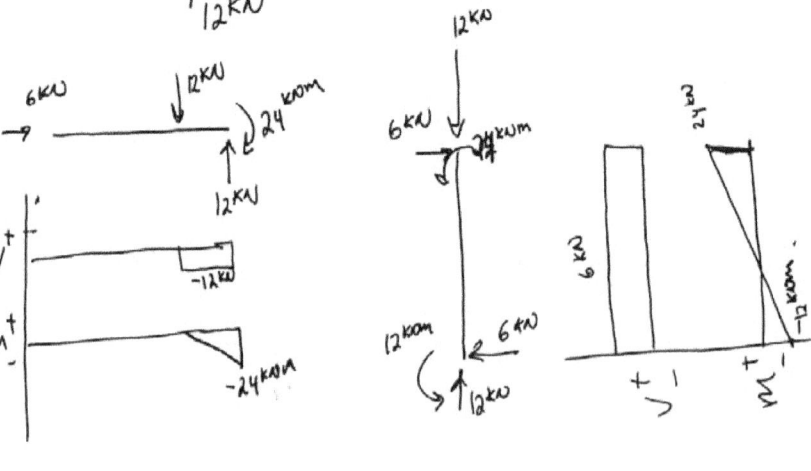

6 kN →   ↓12 kN

← 6 kN
↑ ↺ 12 kNm
↓ 12 kN

<u>Hints</u> - Moment Distribution Method

Distribution Factors

Rollers and pins $\Rightarrow$ DF = $\underline{\underline{1}}$ - when they
are located at the end of a beam or frame.

<u>ie/</u>

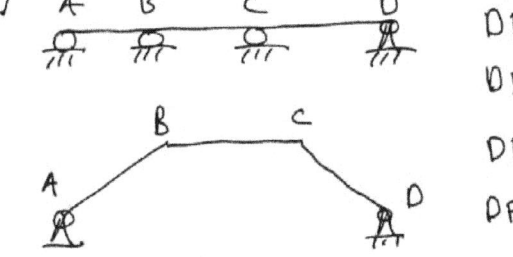

$DF_{AB} = 1$

$DF_{DC} = 1$

$DF_{AB} = 1$

$DF_{DC} = 1$

Fixed supports $\Rightarrow$ DF = $\underline{O}$

$DF_{AB} = \underline{O}$

$DF_{ED} = \underline{O}$

IF you have a cantilever add the moment to the first inside joint

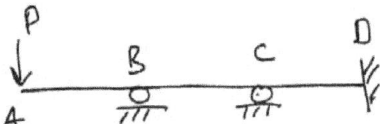

So the model for your chart looks like

B   Fixed end moments

C   D

$DF_{AB} = O$ $DF_{BC} = \underline{1}$ $DF_{DC} = O$

$DF_{BA} = O$

Beam 1

Determine the reactions at the supports using the Moment Distribution Method.

$EI = $ constant
A & C are pins
B is a roller

$$DF_{AB} = \frac{\frac{4EI}{L}}{\frac{4EI}{L} + 0} = \underline{\underline{1}}$$

(No other member attached to A)

$$DF_{CB} = \frac{\frac{4EI}{L}}{\frac{4EI}{L} + 0} = \underline{\underline{1}}$$

(2 memebers attached to C but CD does not add any stiffness because CD can freely rotate)

$$DF_{BA} = \frac{\frac{3EI}{L} \leftarrow \text{(BA)}}{\frac{3EI}{L} + \frac{3EI}{L}}$$

(BA) ↗ L   (BC)

Note: the simplified method is used where the end of beam ends in a pin or roller
$$\frac{3EI}{L}$$

$$DF_{BA} = \frac{\frac{3EI}{10^{ft}}}{\frac{3EI}{10^{ft}} + \frac{3EI}{25^{ft}}} = \underline{\underline{.714}}$$

(DF_{BC} is also equal to $1 - DF_{BA}$ DF's have to add to 1 at a joint)

$$DF_{BC} = \frac{\frac{3EI}{L}}{\frac{3EI}{L} + \frac{3EI}{L}} = \frac{\frac{3EI}{25^{ft}}}{\frac{3EI}{25^{ft}} + \frac{3EI}{10^{ft}}} = \underline{\underline{0.286}}$$

Beam 1

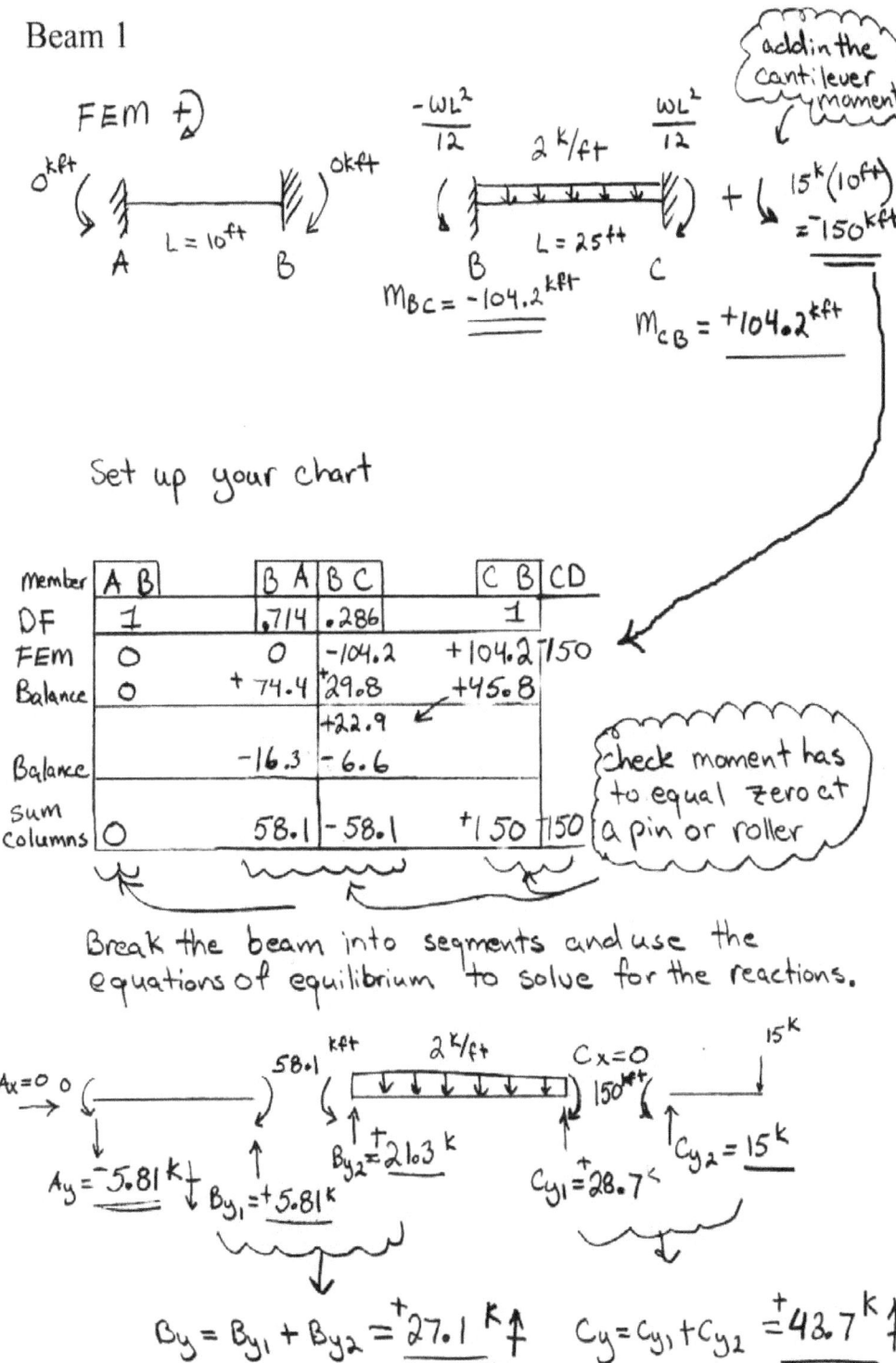

FEM $\tilde{+}$)

0^{kft} A $L = 10^{ft}$ B 0^{kft}

$\dfrac{-WL^2}{12}$ $2^{k/ft}$ $\dfrac{WL^2}{12}$ add in the cantilever moment

B $L = 25^{ft}$ C $+$ $15^k(10ft)$ $= ^-150^{kft}$

$M_{BC} = -104.2^{kft}$ $M_{CB} = +104.2^{kft}$

Set up your chart

Member	A B		B A	B C		C B	CD
DF	1		.714	.286		1	
FEM	0		0	−104.2		+104.2	150
Balance	0		+74.4	+29.8		+45.8	
				+22.9			
Balance			−16.3	−6.6			
Sum Columns	0		58.1	−58.1		+150	150

check moment has to equal zero at a pin or roller

Break the beam into segments and use the equations of equilibrium to solve for the reactions.

$A_x = 0$ 0

58.1^{kft} $2^{k/ft}$ $C_x = 0$ 15^k

150^{kft}

$A_y = ^-5.81^k$ $B_{y_1} = ^+5.81^k$ $B_{y_2} = 21.3^k$ $C_{y_1} = 28.7^<$ $C_{y_2} = 15^k$

$B_y = B_{y_1} + B_{y_2} = ^+27.1^k \uparrow$ $C_y = C_{y_1} + C_{y_2} = ^+43.7^k \uparrow$

Beam 2

$EI = constant$

Determine the support reactions.

$DF_{AB} = \underline{\underline{1}}$ $DF_{DC} = \underline{\underline{0}}$ [factor for a pin support]

$DF_{BA} = \dfrac{\dfrac{3EI}{L} \}}{\dfrac{3EI}{L} + \dfrac{4EI}{L}} = \dfrac{\dfrac{3}{10}}{\dfrac{3}{10} + \dfrac{4}{10}} = \underline{\underline{0.428}}$

[have to add to 1]

$DF_{BC} = 1 - .428 = \underline{\underline{0.572}}$

$DF_{CB} = \dfrac{\dfrac{4EI}{L}}{\dfrac{4EI}{L} + \dfrac{4EI}{L}} = \dfrac{\dfrac{4.}{10}}{\dfrac{4}{10} + \dfrac{4}{15}} = \underline{\underline{0.6}}$

$DF_{CD} = 1 - 0.6 = \underline{\underline{0.4}}$

$\underline{FEM \quad +\circlearrowright}$

$= \dfrac{\cancel{WL^2}}{\cancel{12}} \dfrac{PL}{8}$ $= \dfrac{PL}{8}$ $= \dfrac{-WL^2}{12}$ $= \dfrac{WL^2}{12}$

$= \dfrac{-10^k(10^{ft})}{8}$ $= \dfrac{+10^k(10^{ft})}{8}$ $= \dfrac{-9^{k/ft}\left(10^{ft}\right)^2}{12}$ $= \dfrac{9^{k/ft}\left(10^{ft}\right)^2}{12}$

$= \underline{\underline{-12.5^{kft}}}$ $= \underline{\underline{12.5^{kft}}}$ $= \underline{\underline{-75^{kft}}}$ $= \underline{\underline{75^{kft}}}$

Beam 2 ## FEM cont.

9 k/ft

m_{CD} (beam, L = 15 ft) m_{DC}

L = 15 ft

$$= \frac{-WL^2}{12}$$

$$= \frac{-9(15ft)^2}{12}$$

$$= -168.75^{kft}$$

$$= \frac{WL^2}{12}$$

$$= \frac{9 k/ft (15ft)^2}{12}$$

$$= 168.75^{kft}$$

Chart

	A B		B A	B C		C B	C D		D C
DF	1		0.428	0.572		0.6	0.4		0
FEM	-12.5		12.5	-75		75	-168.75		168.75
	12.5		26.75	35.75		56.25	37.5		0
			6.25	28.125		17.875			18.75
			-14.71	-19.66		-10.725	-7.15		0
				5.36		-9.83			-3.58
			2.29	3.06		5.92	3.93		0
				2.96		1.53			1.97
			-1.27	-1.69		-.92	-.61		0
				-.46		-.85			-.1836
			.20	.26		.51	.34		
				.26		.13			.17
Sum			-.11	-.15		-.08	.05		0
Final moments	0		31.9	-31.9		134.8	-134.8		185.9

Beam 2

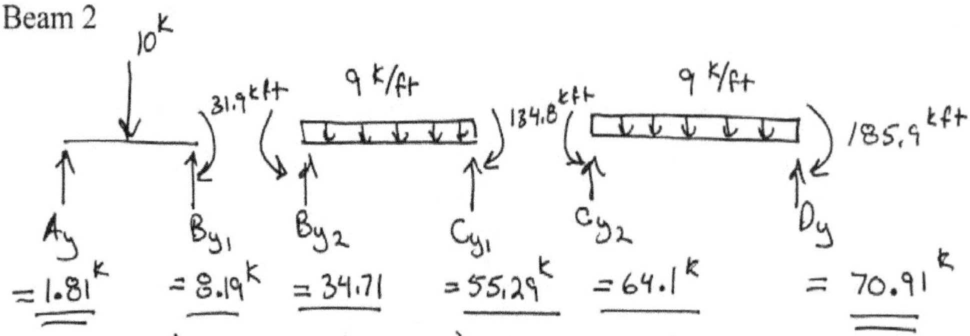

$$A_y = 1.81^k \quad B_y = 42.9^k \quad C_y = 119.4^k \quad D_y = 70.91^k$$

$$D_x = 0^k$$

Solve each segment

Beam 3

$$\underline{Model\ 1}$$

fixed   fixed   fixed

Distribution factors (DF)

$$DF_{AB} = \frac{k}{\xi k} = \frac{k_{AB}}{k_{AB}} = \underline{\underline{1}}$$

$$DF_{BA} = \frac{k}{\xi k} = \frac{k_{BA}}{k_{BA} + k_{BC}}$$

Fixed Ends

$$= \frac{\frac{4EI}{L}}{\frac{4EI}{L} + \frac{4EI}{L}} = \frac{\frac{4EI}{20^{ft}}}{\frac{4EI}{20^{ft}} + \frac{4EI}{15^{ft}}}$$

$$DF_{BA} = \underline{\underline{.429}}$$

$$DF_{BC} = 1 - .429 = \underline{\underline{.571}}$$

$$DF_{CB} = \frac{k}{\xi k} = \frac{k_{CB}}{k_{CB}} = \underline{\underline{1}}$$

$$\underline{Model\ 2}$$

fixed

Distribution Factors (DF)

$$DF_{AB} = \frac{k_{AB}}{k_{AB}} = \underline{\underline{1}}$$

$$DF_{BA} = \frac{k}{\xi k} = \frac{k_{BA}}{k_{BA} + k_{BC}}$$

Pinned Ends

$$= \frac{\frac{3EI}{L}}{\frac{3EI}{L} + \frac{3EI}{L}}$$

$$DF_{BA} = \frac{\frac{3EI}{20^{ft}}}{\frac{3EI}{20^{ft}} + \frac{3EI}{15^{ft}}} = \underline{\underline{.429}}$$

$$DF_{BC} = 1 - .429 = \underline{\underline{.571}}$$

$$DF_{CB} = \frac{k}{\xi k} = \frac{k_{CB}}{k_{CB}} = \underline{\underline{1}}$$

FEM for Both Models

$-\frac{wL^2}{12}$ $+\frac{wL^2}{12}$

$$M_{AB} = -\frac{10^{k/ft}(20^{ft})^2}{12} = \underline{\underline{-333.33}}^{kft} \qquad M_{BC} = \underline{\underline{0}}$$

$$M_{BA} = \frac{10^{k/ft}(20^{ft})^2}{12} = \underline{\underline{333.33}}^{kft} \qquad M_{CB} = \underline{\underline{0}}$$

Beam 3

Model 1

	AB	BA	BC	CB
DF	1	.429	.571	1
FEM	-333.33	333.33		
Bal	333.33	-143	-190.33	
COF	-71.5	166.66	-95.16	
	71.5	-71.499	-95.16	95.16
	-35.75	35.75	47.58	-47.6
	35.75	-35.75	-47.58	47.6
	-17.875	17.87	23.8	-23.8
	17.875	-17.87	-23.8	23.8
	-8.9	8.9	11.9	-11.9
	8.9	-8.9	-11.9	11.9
	-4.45	4.45	5.95	-5.95
	4.45	-4.45	-5.95	5.95
	-2.22	2.22	2.98	-2.98
	2.22	-2.22	2.98	2.98
	-1.11	1.11	1.49	-1.49
	1.11	-1.11	-1.49	1.49
	-.55	.55	.745	-.745
	.55	-.55	-.745	.745
	-.27	.27	.37	-.37
	.27	-.27	-.37	.37
Final Moments	0	285.49	-285.49	0
	M_{AB}	M_{BA}	M_{BC}	M_{CB}

Model 2

	AB	BA	BC	CB
DF	1	.429	.571	1
FEM	-333.33	333.33		
Bal	333.33	-143	-190.33	
COF		166.66		
		-71.499	-95.16	
	0	285.49	-285.49	0
	M_{AB}	M_{BA}	M_{BC}	M_{CB}

Beam 3

$$10^k/ft$$

$A_x \rightarrow$ 285.49

$A_y \uparrow$ B_{y_1} B_{y_2} $C_y \uparrow$

$\overset{+}{\circlearrowleft}$ $\Sigma M_B = 0$

$$10^k/ft(20ft)(10ft) - 285.49 + A_y(20ft) = 0$$

$$A_y = \underline{\underline{85.73^k}} \uparrow$$

$\uparrow \Sigma F_y = 0$

$$-10^k/ft(20ft) + 85.73^k + B_{y_1} = 0$$

$$B_{y_1} = \underline{114.3^k}$$

$\overset{+}{\circlearrowleft}$ $\Sigma M_B = 0$

$$C_y(15ft) + 285.49 = 0$$

$$C_y = \underline{\underline{19.1^k}} \downarrow$$

$\uparrow \Sigma F_y = 0$

$$-19.1^k + B_{y_2} = 0$$

$$B_{y_2} = \underline{\underline{19.1^k}}$$

$$B_y = B_{y_1} + B_2$$

$$B_y = 114.3^k + 19.1^k = \underline{\underline{133.33^k}} \uparrow$$

Final structure

$$10^k/ft$$

$0^k \rightarrow$

85.73^k 133.33^k $\downarrow 19.1^k$

$20ft$ $15ft$

Beam 4

Draw the shear and moment diagram for the beam. $EI = $ constant

1) Distribution Factors (DF)

$DF_{AB} = \underline{0}$ (Always for a fixed end)

$DF_{CB} = \underline{1}$ (Always for a pin or roller end)

$$D_{FBA} = \frac{k_{BA}}{k_{BA} + k_{BC}} = \cfrac{\frac{4EI}{16ft}}{\frac{4EI}{16ft} + \frac{3EI}{24ft}} = \underline{2/3}$$

$$DF_{BC} = 1 - 2/3 = \underline{1/3}$$

2) Fixed End Moments (FEM) $+\circlearrowleft$

M_{AB}

$= \frac{-WL^2}{12} = \frac{-6^{k}/ft(16ft)^2}{12} = \underline{-128^{kft}}$

M_{BA}

$= \frac{+WL^2}{12} = \underline{128^{kft}}$

M_{BC}

$= \frac{-WL^2}{12} = \underline{-288^{kft}}$

M_{CB}

$= \frac{+WL^2}{12} = \underline{+288^{kft}}$

$+$

M_{BC} $a = 8ft$ $b = 16ft$

$= \frac{-Pb^2a}{L^2} = \underline{-71.1^{kft}}$

M_{CB}

$= \frac{Pa^2b}{L^2} = \underline{35.6^{kft}}$

Beam 4

<u>Chart</u>

	A B		BA	BC		CB
DF	O		⅔	⅓		1
FEM	-128		128	-359.1		323.6
	O		154.1	77		-323.6
	77.2			-161.8		
	O		107.9	53.9		
	53.95					
Final	3.15 kft		390 kft	-390 kft		O

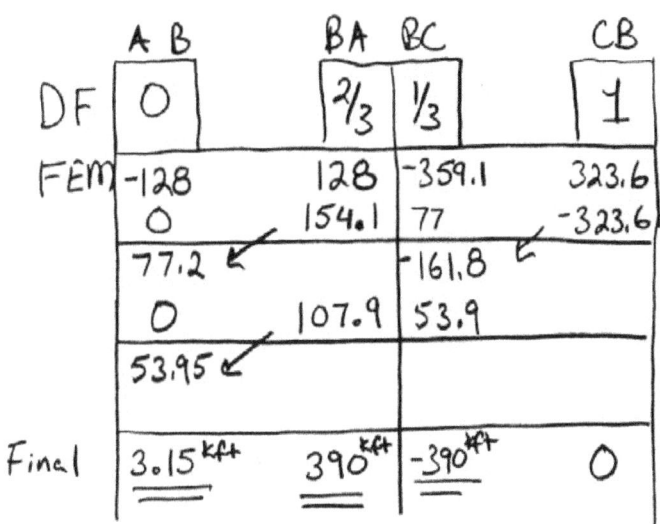

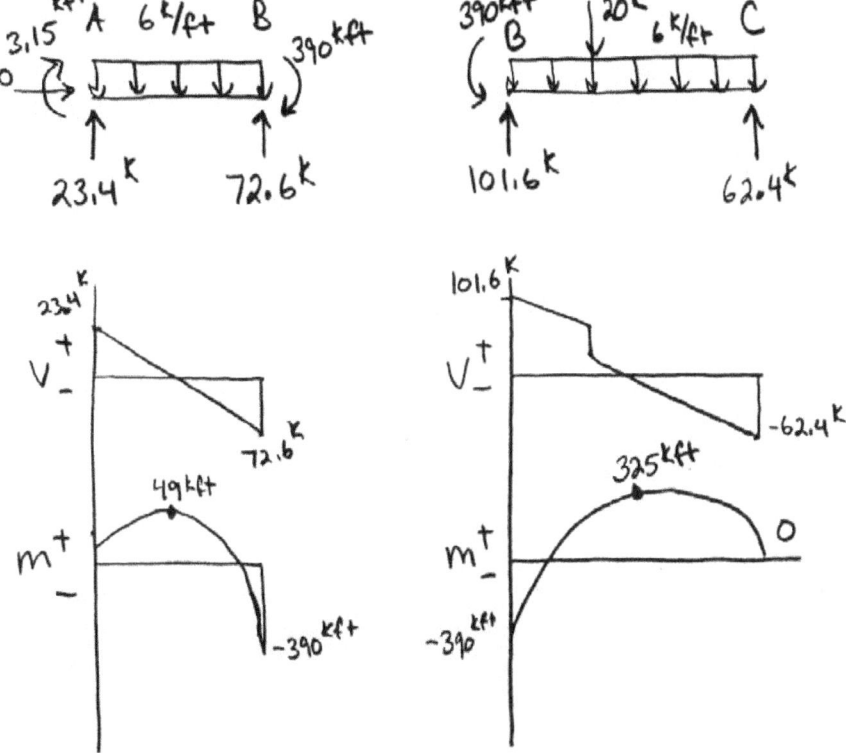

A $\downarrow$ 10^k B 4 ^{k}/ft C

Draw the shear and
moment Diagram for the beam
E = constant
I = 1000 in^4 for AB
I = 2000 in^4 for BC

5 ft   5 ft   15 ft

1) Distribution factors (DF)

$$DF_{AB} = 1 \qquad DF_{BA} = \frac{k_{BA}}{k_{BA} + k_{BC}} = \frac{\dfrac{3EI}{10ft}}{\dfrac{3EI}{10ft} + \dfrac{4EI(2)}{15ft}} = .36$$

$$DF_{CB} = 0$$

$$DF_{BC} = 1 - .36 = .64$$

2) Fixed end Moments (FEM)

M_{AB} 4 ^{k}/ft M_{BA}

10 ft

$$= \frac{-WL^2}{12}$$

$$= \frac{-4^k/ft(10^{ft})^2}{12}$$

$$= -33\tfrac{1}{3} ^{kft}$$

$$= \frac{WL^2}{12}$$

$$= 33\tfrac{1}{3} ^{kft}$$

M_{BC} 4 ^{k}/ft M_{CB}

15 ft

$$= \frac{-WL^2}{12}$$

$$= \frac{-4^k/ft(15^{ft})^2}{12}$$

$$= -75^{kft}$$

$$= \frac{WL^2}{12}$$

$$= 75^{kft}$$

+

M_{AB} $\downarrow$ 10^k M_{BA}

10 ft

$$= \frac{-PL}{8}$$

$$= \frac{-10^k(10^{ft})}{8}$$

$$= -12.5 ^{kft}$$

$$= \frac{PL}{8}$$

$$= 12.5 ^{kft}$$

Beam 5

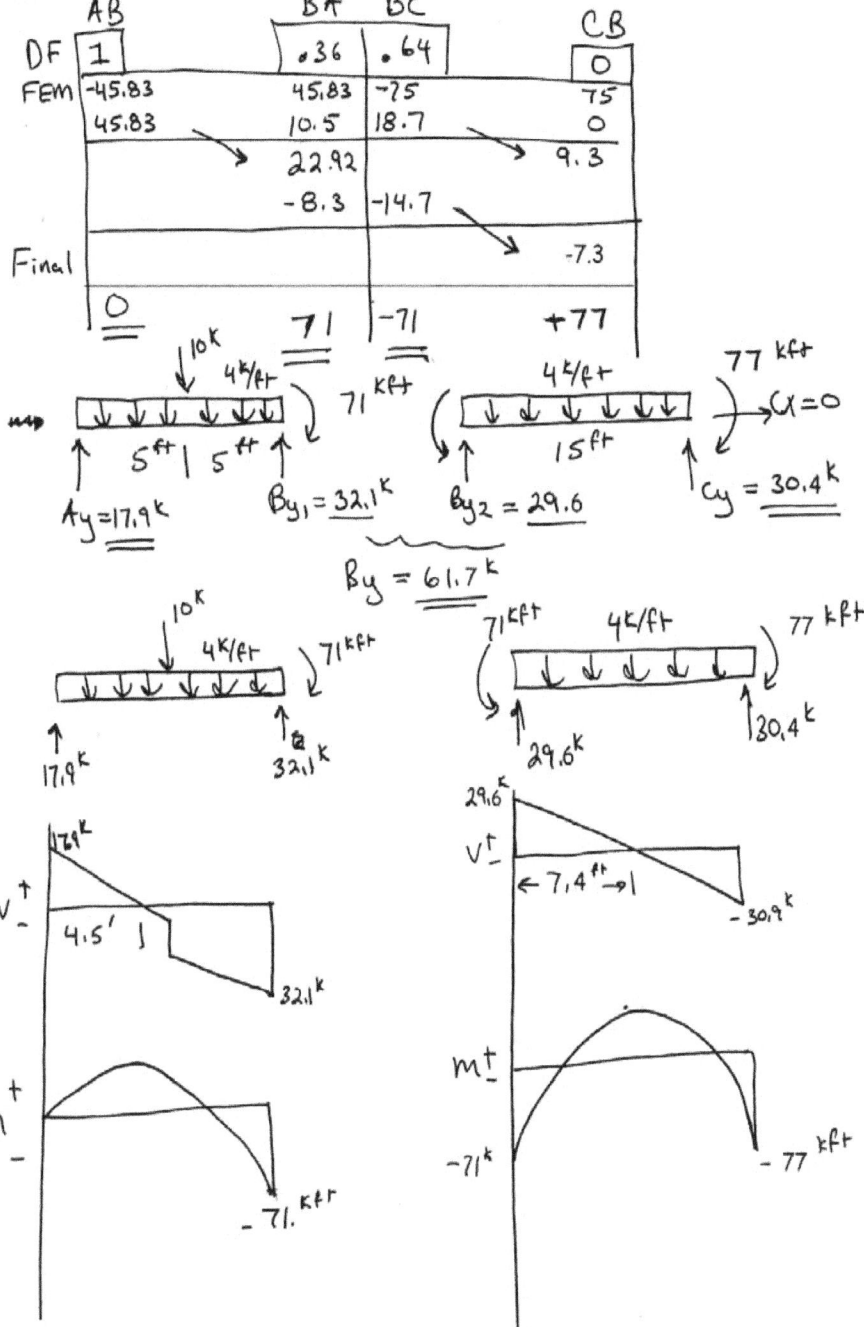

	AB	BA	BC	CB
DF	1	.36	.64	0
FEM	-45.83	45.83	-75	75
	45.83	10.5	18.7	0
		22.92		9.3
		-8.3	-14.7	
Final				-7.3
	0	71	-71	+77

$A_y = 17.9^k$

$B_{y_1} = 32.1^k$

$B_{y_2} = 29.6$

$C_y = 30.4^k$

$B_y = 61.7^k$

$\Sigma X = 0$

17.9^k 32.1^k

29.6^k 30.4^k

17.9^k

$4.5'$

32.1^k

$-71. \, kft$

29.6^k

$\leftarrow 7.4^{ft} \rightarrow$

-30.9^k

-71^k

$-77 \, kft$

1) $\underline{DF}$

$DF_{AB} = 1 \quad DF_{CB} = 0$

$$DF_{BA} = \frac{k_{BA}}{k_{BA} + k_{BC}} = \frac{\dfrac{3EI}{10ft}}{\dfrac{3EI}{10ft} + \dfrac{4EI}{15ft}} = .529$$

$DF_{BC} = 1 - .529 = \underline{.471}$

2) FEM

MAB $4^k/ft$ MBA MBC $4^k/ft$ M_CB

$L = 10ft$ $L = 15ft$

$= \dfrac{-WL^2}{12}$ $= \dfrac{WL^2}{12}$ $= \dfrac{-WL^2}{12}$ $= \dfrac{WL^2}{12}$

$= \dfrac{-4^k/ft(10^{ft})^2}{12}$ $= \underline{33.33^{kft}}$ $= \dfrac{-4^k/ft(15^{ft})^2}{12}$ $= \underline{75^{kft}}$

$= \underline{-33.33}^{kft}$ $= \dfrac{-75^{kft}}{12}$

$= \underline{-75^{kft}}$

$+$

MAB 10^k MBA

$5ft$ $5ft$

$= \dfrac{PL}{8}$ $= \dfrac{-PL}{8}$

$= \dfrac{10^k(10^{ft})}{8}$ $= \underline{-12.5^k}$

$= \underline{12.5^k}$

$M_{AB} = -33.33 + 12.5 = \underline{\underline{-20.83^{kft}}}$

$M_{BA} = -12.5 + 33.33 = \underline{\underline{20.83^{kft}}}$

Beam 6

	AB	BA	BC	CB
DF	1	.529	.471	0
FEM	20.83	20.83	-75	75
	20.83	28.656	25.514	0
		10.4		12.757
		-5.5	-4.9	
				-2.45
Final Moments	0	54.4	-54.4	85.3

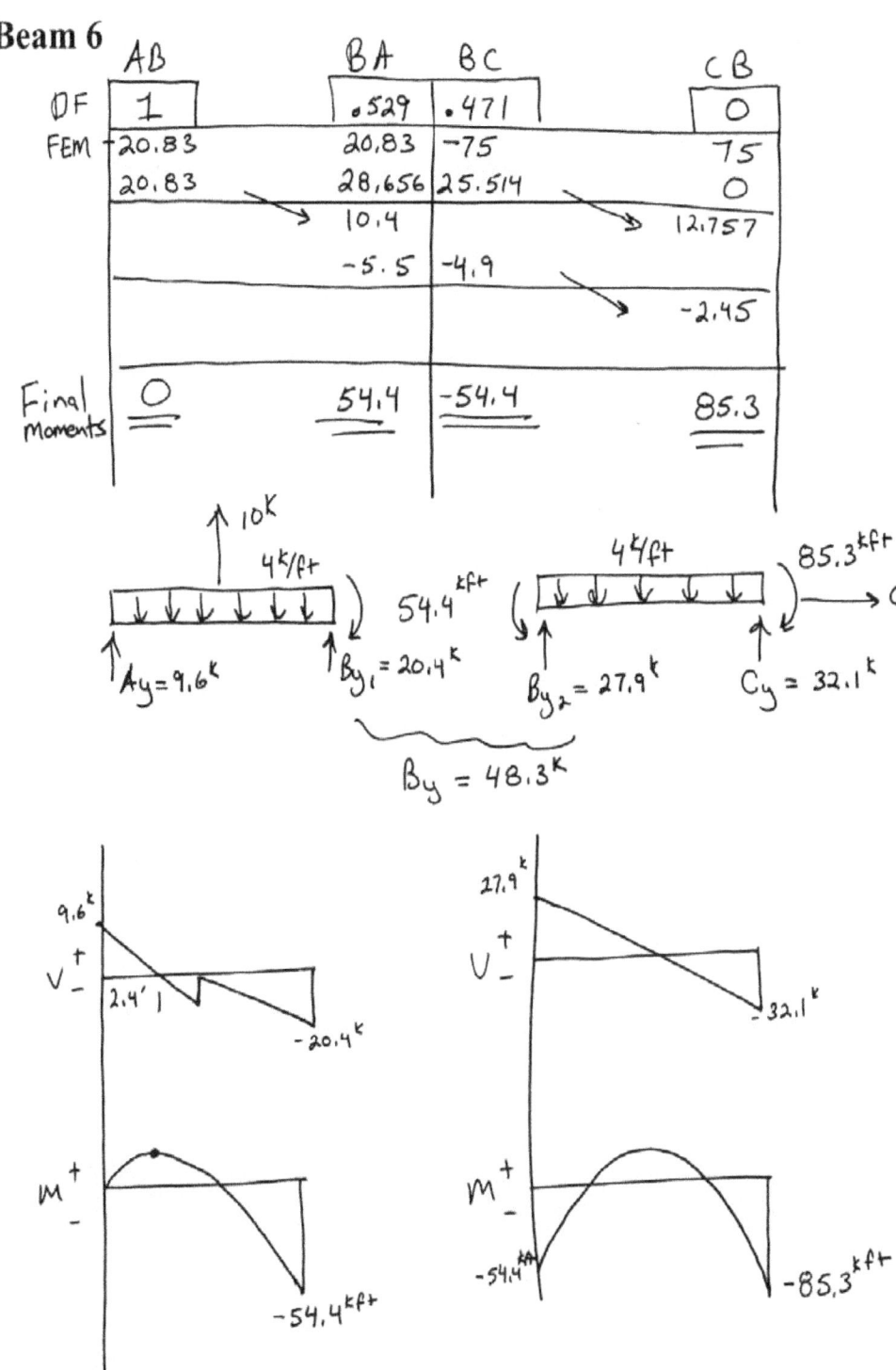

Beam 7

A   B 3^k/ft   C   7^k/ft   D   E Determine the support reactions for the beam EI = constant.

15ft   20 ft   25ft   10ft

1) Distribution Factors (DF)

$$DF_{AB} = \underline{0} \text{ (fixed end)} \quad DF_{ED} = \underline{0} \text{ (fixed end)}$$

$$DF_{BA} = \frac{k_{BA}}{k_{BA}+k_{BC}} = \frac{\frac{4EI}{15ft}}{\frac{4EI}{15ft} + \frac{4EI}{20ft}} = \underline{.571} \quad DF_{BC} = 1-.571 = \underline{.429}$$

$$DF_{CB} = \frac{k_{CB}}{k_{CB}+k_{CD}} = \frac{\frac{4EI}{20ft}}{\frac{4EI}{20ft} + \frac{4EI}{25ft}} = \underline{.56} \quad DF_{CD} = 1-.56 = \underline{.44}$$

$$DF_{DC} = \frac{k_{DC}}{k_{DC}+k_{DE}} = \frac{\frac{4EI}{25ft}}{\frac{4EI}{25ft} + \frac{4EI}{10ft}} = \underline{.29} \quad DF_{DE} = 1-.29 = \underline{.71}$$

2) FEM $\overset{+}{\downarrow}$

M_{AB} 3^k/ft M_{BA} M_{BC} 3^k/ft M_{CB} M_{CD} 7^k/ft M_{DC} M_{DE} 7^k/ft M_{ED}

15ft   20 ft   25ft   10ft

$= -56.25^{kft}$ $= 56.25^{kft}$ $= -100^{kft}$ $= 100^{kft}$ $= -364.6^{kft}$ $= 364.6$^{kft} $= -58.3$^{kft} $= 58.3$^{kft}

using

M w m

$$= \frac{-wl^2}{12} \qquad = \frac{wl^2}{12}$$

Beam 7

	AB	BA	BC	CB	CD	DC	DE	ED
DF	0	0.571	0.429	0.56	0.44	0.29	0.71	0
FEM	-56.25	56.25	-100.00	100.00	-364.60	364.60	-58.30	58.30
Balance	0.00	24.98	18.77	148.18	116.42	-88.83	-217.47	0.00
COF	12.49	0.00	74.09	9.38	-44.41	58.21	0.00	-108.74
Balance	0.00	-42.30	-31.78	19.62	15.41	-16.88	-41.33	0.00
COF	-21.15	0.00	9.81	-15.89	-8.44	7.71	0.00	-20.67
Balance	0.00	-5.60	-4.21	13.63	10.71	-2.23	-5.47	0.00
COF	-2.80	0.00	6.81	-2.10	-1.12	5.35	0.00	-2.74
Balance	0.00	-3.89	-2.92	1.80	1.42	-1.55	-3.80	0.00
COF	-1.95	0.00	0.90	-1.46	-0.78	0.71	0.00	-1.90
Balance	0.00	-0.52	-0.39	1.25	0.98	-0.21	-0.50	0.00
COF	-0.26	0.00	0.63	-0.19	-0.10	0.49	0.00	-0.25
Balance	0.00	-0.36	-0.27	0.17	0.13	-0.14	-0.35	0.00
COF	-0.18	0.00	0.08	-0.13	-0.07	0.07	0.00	-0.17
Balance	0.00	-0.05	-0.04	0.12	0.09	-0.02	-0.05	0.00
COF	-0.02	0.00	0.06	-0.02	-0.01	0.05	0.00	-0.02
Balance	0.00	-0.03	-0.02	0.02	0.01	-0.01	-0.03	0.00
COF	-0.02	0.00	0.01	-0.01	-0.01	0.01	0.00	-0.02
Balance	0.00	0.00	0.00	0.01	0.01	0.00	0.00	0.00
COF	0.00	0.00	0.01	0.00	0.00	0.00	0.00	0.00
Balance	0.00	0.00	0.00	0.00	0.00	0.00	0.00	0.00
COF	0.00	0.00	0.00	0.00	0.00	0.00	0.00	0.00
Balance	0.00	0.00	0.00	0.00	0.00	0.00	0.00	0.00
Sum	-70.1	28.5	-28.5	274.4	-274.4	327.3	-327.3	-76.2

$A_x = \underline{0^k}$ $m_A = \underline{70.1^{kft}} \curvearrowright$

$A_y = \underline{25.2^k} \uparrow$ $m_D = \underline{76.2^{kft}} \curvearrowright$

$B_y = \underline{37.6^k} \uparrow$

$C_y = \underline{165^k} \uparrow$

$D_y = \underline{-5.3^k} \downarrow$

$D_x = \underline{0^k}$

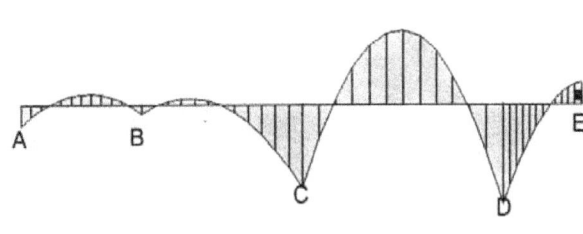

A $5^{k}/ft$ 15^{k} D

Draw the shear and Moment Diagram for the beam. EI = constant

20ft | 15ft | 5ft | 5ft
B | C

1) Distribution Factors (DF)

$DF_{AB} = \underline{0}$ $DF_{DC} = \underline{1}$

$$DF_{BA} = \frac{k_{BA}}{k_{BA} + k_{BC}} = \frac{\frac{4EI}{20^{ft}}}{\frac{4EI}{20^{ft}} + \frac{4EI}{15^{ft}}} = \underline{.429}$$

$DF_{BC} = 1 - .429 = \underline{.571}$

$$DF_{CB} = \frac{k_{CB}}{k_{CB} + k_{CD}} = \frac{\frac{4EI}{20^{ft}}}{\frac{4EI}{20^{ft}} + \frac{3EI}{10^{ft}}} = \underline{.4}$$ $DF_{CD} = 1 - .4 = \underline{.6}$

2) FEM ↻ (+)

M_{AB} $5^{k}/ft$ M_{BA} M_{BC} $5^{k}/ft$ M_{CB} M_{CD} 15^{k} M_{DC}

L = 20ft   L = 15ft   L = 10ft

$\frac{-WL^2}{12}$ $\frac{WL^2}{12}$ $\frac{-WL^2}{12}$ $\frac{WL^2}{12}$ $\frac{PL}{8}$ $\frac{-PL}{8}$

$= -166.67^{kft}$ $= 166.67^{kft}$ $= -93.75^{kft}$ $= 93.75^{kft}$ $= 18.75^{kft}$ $= -18 kft$

Beam 8

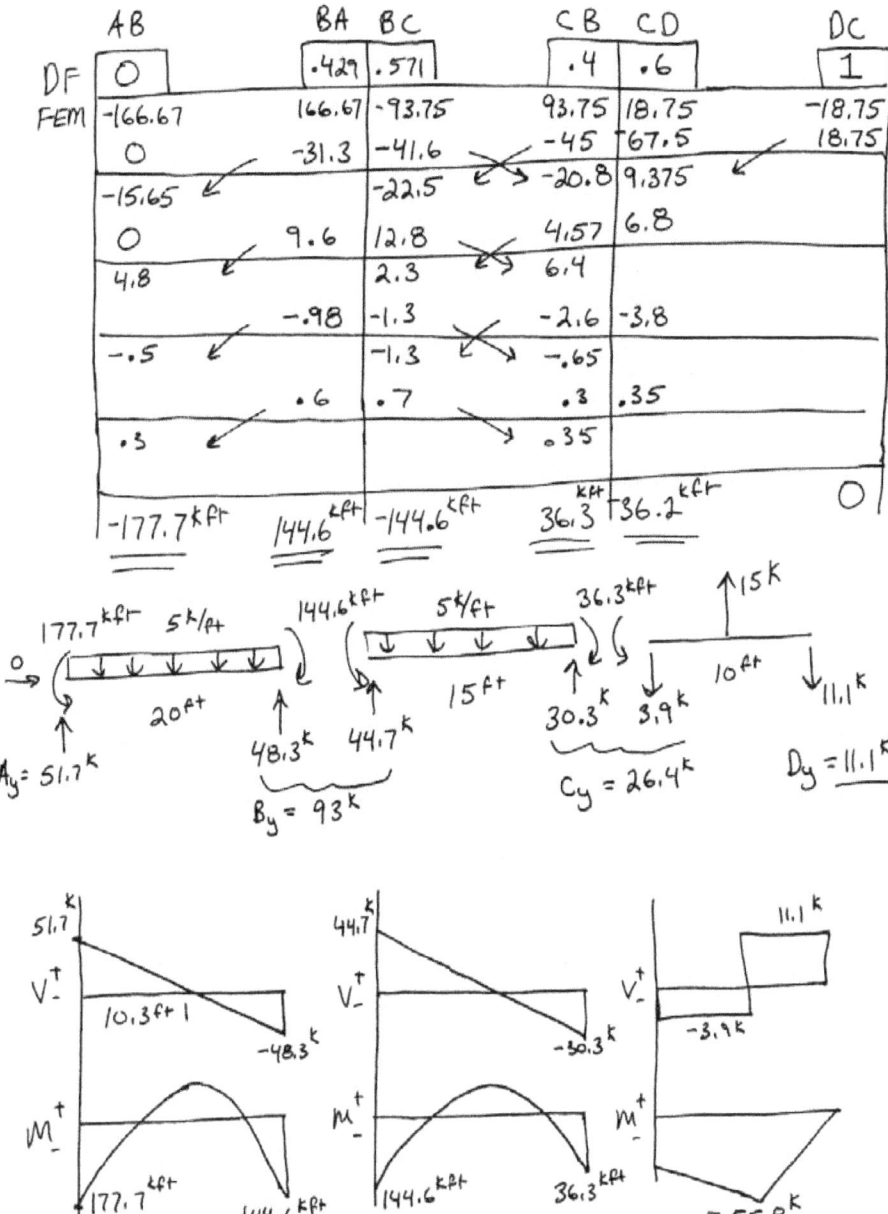

	AB		BA	BC		CB	CD		DC
DF	0		.429	.571		.4	.6		1
FEM	-166.67		166.67	-93.75		93.75	18.75		-18.75
	0		-31.3	-41.6		-45	67.5		18.75
	-15.65			-22.5		-20.8	9.375		
	0		9.6	12.8		4.57	6.8		
	4.8			2.3		6.4			
			-.98	-1.3		-2.6	-3.8		
	-.5			-1.3		-.65			
			.6	.7		.3	.35		
	.3					.35			
	-177.7^{kft}		144.6^{kft}	-144.6^{kft}		36.3^{kft}	36.2^{kft}		0

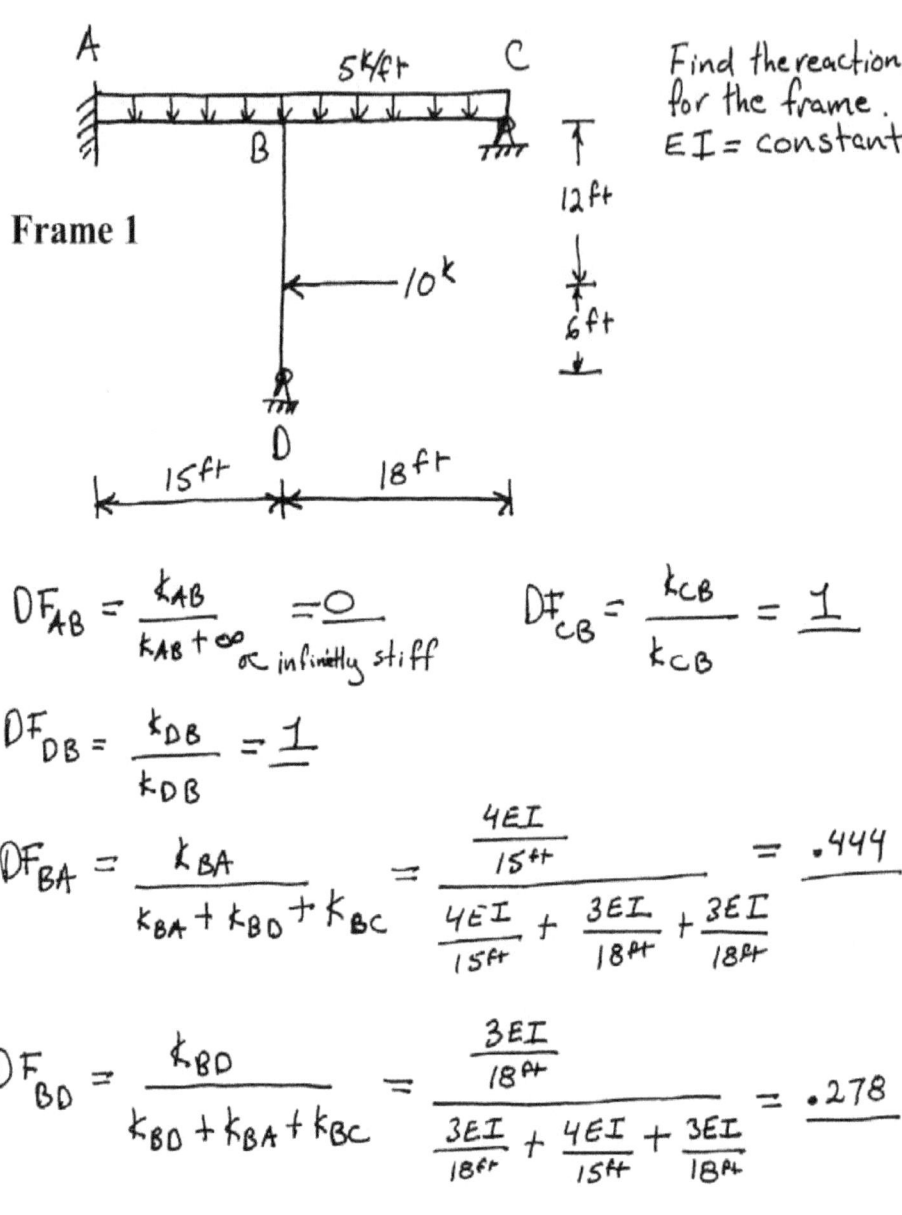

Frame 1

Find the reactions for the frame. $EI = $ constant.

$$DF_{AB} = \frac{k_{AB}}{k_{AB} + \infty} = 0$$
∞ infinitly stiff

$$DF_{CB} = \frac{k_{CB}}{k_{CB}} = \underline{1}$$

$$DF_{DB} = \frac{k_{DB}}{k_{DB}} = \underline{1}$$

$$DF_{BA} = \frac{k_{BA}}{k_{BA} + k_{BD} + k_{BC}} = \frac{\frac{4EI}{15ft}}{\frac{4EI}{15ft} + \frac{3EI}{18ft} + \frac{3EI}{18ft}} = \underline{.444}$$

$$DF_{BD} = \frac{k_{BD}}{k_{BD} + k_{BA} + k_{BC}} = \frac{\frac{3EI}{18ft}}{\frac{3EI}{18ft} + \frac{4EI}{15ft} + \frac{3EI}{18ft}} = \underline{.278}$$

$$DF_{BC} = 1 - .444 - .278 = \underline{.278}$$

FEM

A   5^k/ft   B

m_{AB} L=15ft m_{BA}

$= \dfrac{-wL^2}{12}$

$= \dfrac{-wL^2}{12} = 93.75^{kft}$

$= -\dfrac{5^k/ft\,(15ft)^2}{12} = \dfrac{-93.75^{kft}}$

B   5^k/ft   C

m_{BC} L = 18ft m_{CB}

$= \dfrac{-wL^2}{12}$

$= \dfrac{wL^2}{12}$

$= -\dfrac{5^k/ft\,(18ft)^2}{12}$

$= 135^{kft}$

$= -135^{kft}$

$m_{BD} = \dfrac{-Pb^2a}{L^2}$

$a = 12ft$

$L = 18ft$

$b = 6\,ft$

10^k

$= -\dfrac{10^k(6^{ft})^2(12^{ft})}{(18ft)^2} = -13.33^{kft}$

$m_{DB} = \dfrac{Pa^2b}{L^2} = \dfrac{10^k(12^{ft})^2(6^{ft})}{(18ft)^2} = 26.67^{kft}$

Member	AB		BA	BD	BC		CB		DB	
DF	0		.444	.278	.278		1		1	
FEM	-93.75		93.75	-13.33	-135		135		26.67	toBD
Balance	0		24.23	15.17	15.17		-135		-26167	↑
CoF	12.12			-13.33	-67.5					
Balance			35.89	22.47	22.47					
CoF	17.95									
Sum	↑63.68^{kft}		153.87^{kft}	10.98^{kft}	-164.86^{kft}					

63.68^{kft}

$5^{k}/ft$

A_x B_{x_1} 153.87^{kft}

$\uparrow A_y$ ① $\uparrow B_{y_1}$

164.86^{kft} $5^{k}/ft$

B_{x_2} C_x

$\uparrow B_{y_2}$ ② $\uparrow C_y$

B_y

B_x 10.98^{kft}

③

10^k

D_x

$\uparrow D_y$

Sample Calculations

Member ①

$f)\ \Sigma M_B = 0$

$$63.68^{kft} - 153.87^{kft} - A_y(15^{ft}) + 5^{k/ft}(15^{ft})(7.5^{ft}) = 0$$

$$A_y = 31.49^k$$

$\uparrow \Sigma F_y = 0$

$$31.49^k - 5^{k/ft}(15^{ft}) + B_{y_1} = 0$$

$$B_{y_1} = 43.51^k$$

Member ②

$f)\ \Sigma M_B = 0$

$$164.86^{kft} - 5^{k/ft}(18^{ft})(9^{ft}) + C_y(18^{ft}) = 0$$

$$C_y = 35.84^k$$

$\uparrow \Sigma F_y = 0$

$$B_{y_2} - 5^{k/ft}(18^{ft}) + 35.84^k = 0$$

$$B_{y_2} = 54.16^k$$

Overall

$$B_y = D_y = B_{y_1} + B_{y_2} = 43.51^k + 54.16^k = 97.67^k \uparrow$$

$$A_x = 1.4^k \rightarrow \quad D_x = 7.4^k \rightarrow \quad C_x = 1.2^k \rightarrow$$

Frame 2

Find the joint moments
EI = Constant
(fixed joint at B)

1) DF

$DF_{AB} = \underline{1}$

$DF_{DB} = \underline{1}$

$DF_{CB} = \underline{1}$

$DF_{BA} = \dfrac{k_{BA}}{k_{BA} + k_{BD} + k_{BC}} = \dfrac{\frac{3EI}{10ft}}{\frac{3EI}{10ft} + \frac{3EI}{30ft} + \frac{3EI}{10ft}} = \underline{.429}$

$DF_{BC} = \dfrac{k_{BC}}{k_{BA} + k_{BD} + k_{BC}} = \dfrac{\frac{3EI}{10ft}}{\frac{3EI}{10ft} + \frac{3EI}{30ft} + \frac{3EI}{10ft}} = \underline{.429}$

$DF_{BD} = 1 - .429 - .429 = \underline{.142}$

2) FEM

$M_{BD} = \dfrac{-PL}{8} = \dfrac{-25^k (30ft)}{8} = \underline{-93.75^{kft}}$

$M_{DB} = \dfrac{PL}{8} = \dfrac{25^k (30ft)}{8} = \underline{93.75^{kft}}$

Chart   add $+200^{kft}$ Frame 2
                                   to -93.75^{kft} and balance

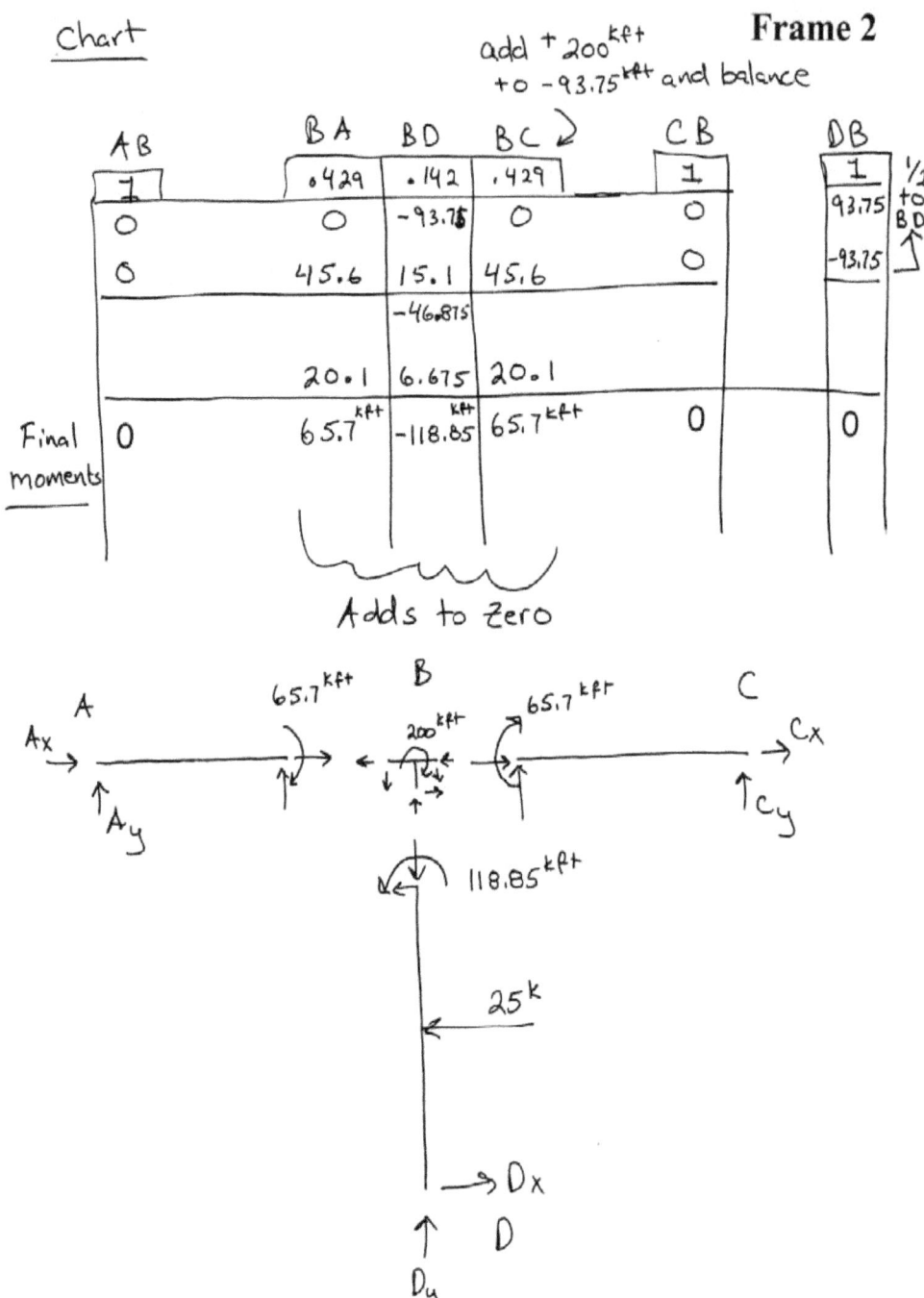

AB	BA	BD	BC	CB	DB	
1	.429	.142	.429	1	1	½
O		-93.75	O	O	93.75	to BD
O	45.6	15.1	45.6	O	-93.75	
		-46.875				
	20.1	6.675	20.1			
Final moments 0	65.7^{kft}	-118.85^{kft}	65.7^{kft}	0	0	

Adds to zero

65.7^{kft} B 65.7^{kft} C

A 200^{kft}
A_x → C_x →
  ↑A_y ↑C_y

                  118.85^{kft}

                  25^k

              → D_x
           ↑   D
          D_y

Frame 3

Find the reactions for the Frame.
EI = constant.
20^k (Frame subject to Sidesway)

But use "Q" Frame instead of εR Frame

M_R frame moments ← Restraint

So $M = M_R + \dfrac{R}{Q} M_Q$

$\underset{force}{C}$ ↖ M_Q frame Moments

1) Distribution Factors (DF)

$DF_{AB} = \underline{0}$ $DF_{DC} = \underline{1}$

$DF_{BA} = \dfrac{k_{BA}}{k_{BA} + k_{BC}} = \dfrac{\frac{4EI}{15ft}}{\frac{4EI}{15ft} + \frac{4EI}{15ft}} = \underline{.5}$ ⎫
                                                                                  ⎬ have to add
$DF_{BC} = \dfrac{k_{BC}}{k_{BC} + k_{BA}} = 1 - .5 = \underline{.5}$ ⎭ to One.

$DF_{CB} = \dfrac{k_{CB}}{k_{CB} + k_{CD}} = \dfrac{\frac{4EI}{15ft}}{\frac{4EI}{15ft} + \frac{3EI}{20ft}} = \underline{.64}$

$DF_{CD} = 1 - .64 = \underline{.36}$

2) Fixed End Moments (FEM).   **Frame 3**

$$M_{CD} = \frac{-PL}{8} = \frac{20^k(20ft)}{8} = -50^k$$

C ⟲📐

← 20^k

O ⟲📐 $M_{DC} = \frac{PL}{8} = \frac{20^k(20ft)}{8} = 50^k$

chart 1

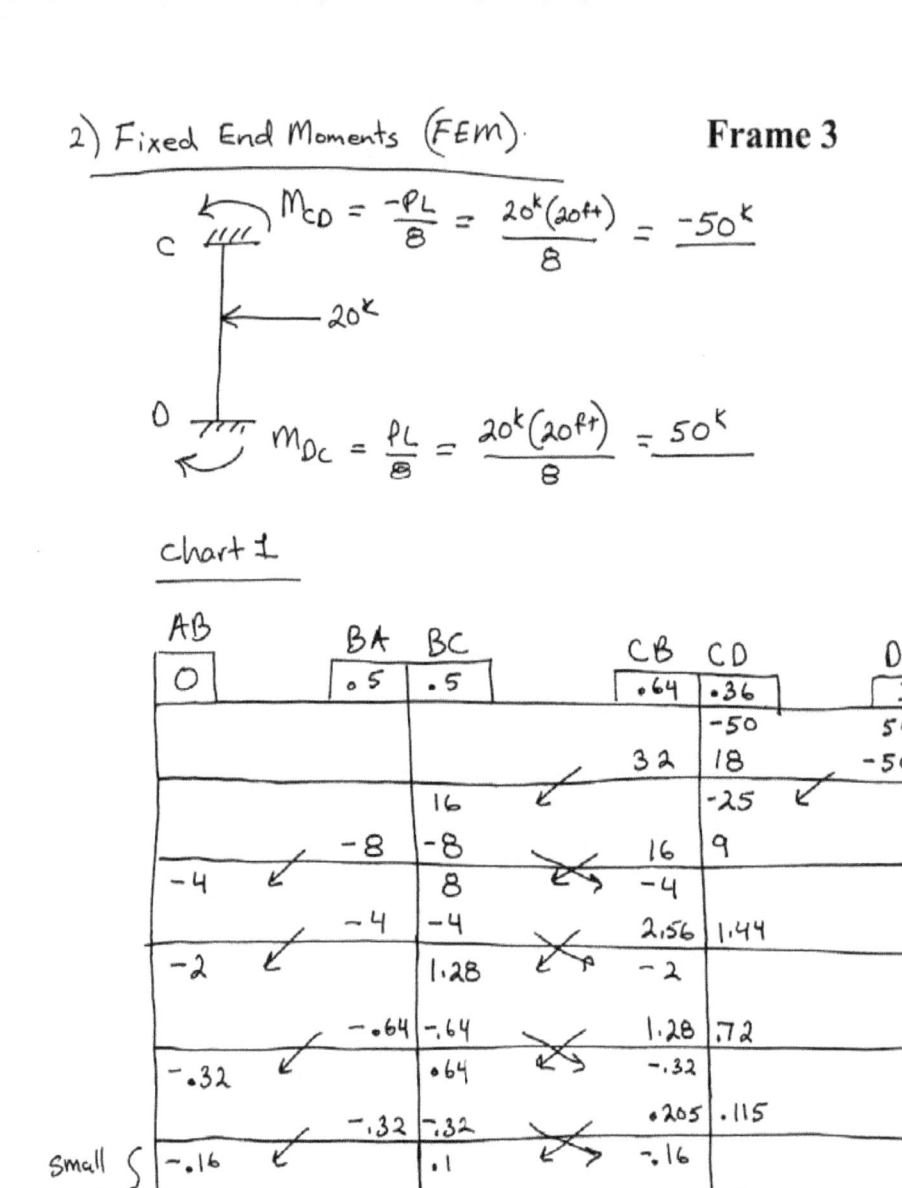

AB		BA	BC		CB	CD		DC
O		.5	.5		.64	.36		1
						-50		50
					32	18		-50
			16			-25		
		-8	-8		16	9		
-4			8		-4			
		-4	-4		2.56	1.44		
-2			1.28		-2			
		-.64	-.64		1.28	.72		
-.32			.64		-.32			
		-.32	-.32		.205	.115		
-.16			.1		-.16			
		-.05	-.05		.1	.06		

Small enough {

M_R |-6.48^{kft} | -13^{kft} | 13^{kft} | 45.7^{kft} | -45.7^{kft} | O |

R 13^{kft} ⟲
1.3^k → ⟳ 1.3^k ↑   B

45.7^{kft} ⟲
12.3^k
↑ C → 12.3^k

← 20^k

A 1.3^k ← ⟳
↑ 6.48^{kft}

7.7^k →
↑ D

$R = 1.3^k + 12.3^k$

$R = 13.6^k$

AB	BA	BC	CB	CD	DC
0	.5	.5	.64	.36	
50	50				
0				28.125	28.125
	-25	-25	-18	-10.125	-28.125
-12.5		-9	-12.5	-14.1	
0	4.5	4.5	17	9.6	
2.25		8.5	2.25		
	-4.25	-4.25	~1.44	-.81	
-2.125		-.7	~2.125		
	.35	.35	1.36	.765	
.18		.68	.18		
	-.34	.34	-.11	-.06	
37.8 $^{k ft}$	25.3 $^{k ft}$	-25.3 $^{k ft}$	-13.4 $^{k ft}$	13.4 $^{k ft}$	0

Small enough to stop

m_Q

$$Q = 4.21^k + .67^k$$

$$Q = 4.88^k \leftarrow$$

$$M = M_R + \frac{R}{Q} M_Q$$

Final Moments — use to Find Reactions

$$M_{AB} = -6.48 + \left(\frac{13.6}{4.88}\right)(37.8) = 98^{k ft}$$

$$m_{BA} = -13 + \left(\frac{13.6}{4.88}\right)(25.3) = 57.5^{k ft} \qquad m_{BC} = -57.5^{k ft}$$

$$m_{CB} = 45.7 + \left(\frac{13.6}{4.88}\right)(-13.4) = 8.4^{k ft} \qquad m_{CD} = -8.4^{k ft}$$

$$m_{DC} = 0$$

Q → |Δ| |Δ|

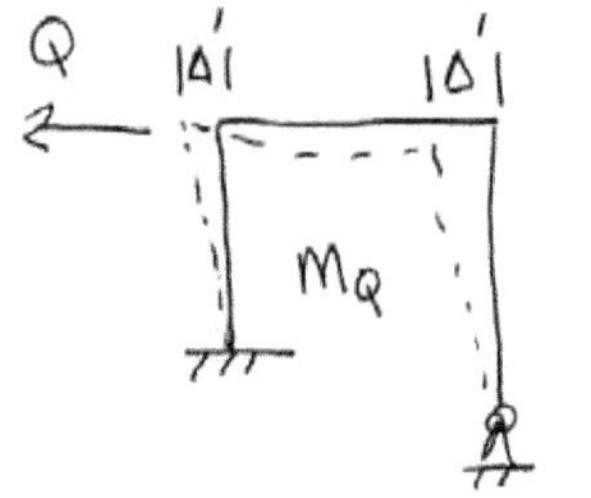

1) Distribution Factors are the Same for the M_Q frame.

M_Q

2) FEM

$m_{BA} = \dfrac{6EI\Delta'}{L^2}$

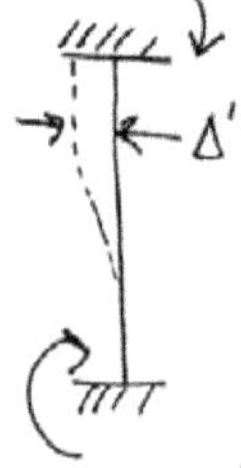

$m_{CD} = \dfrac{6EI\Delta'}{L^2}$

→ ←Δ'

←Δ'

$M_{AB} = \dfrac{6EI\Delta'}{L^2}$

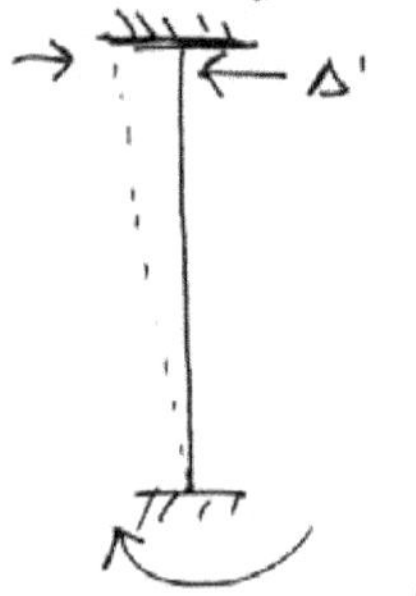

$M_{DC} = \dfrac{6EI\Delta'}{L^2}$

Assume

$$m_{AB} = m_{BA} = \frac{6EI\,\Delta'}{L^2} = \underline{50^{kft}}$$

then $\Delta' = \dfrac{L^2(50^{kft})}{6EI} = \dfrac{(15ft)^2(50^{kft})}{6EI} = \underline{\dfrac{1875}{EI}}$

Plug in to

$$m_{CD} = \frac{6EI\,\Delta'}{L^2} = \frac{6EI}{(20ft)^2}\left(\frac{1815}{EI}\right) = \underline{28.125^{kft}}$$

$$m_{DC} = \underline{28.125^{kft}}$$

Start M_Q chart

www.ingramcontent.com/pod-product-compliance
Lightning Source LLC
Chambersburg PA
CBHW080708190526

45169CB00006B/2288